1,000,000 Books

are available to read at

www.ForgottenBooks.com

Read online
Download PDF
Purchase in print

ISBN 978-0-282-57708-7
PIBN 10857397

This book is a reproduction of an important historical work. Forgotten Books uses state-of-the-art technology to digitally reconstruct the work, preserving the original format whilst repairing imperfections present in the aged copy. In rare cases, an imperfection in the original, such as a blemish or missing page, may be replicated in our edition. We do, however, repair the vast majority of imperfections successfully; any imperfections that remain are intentionally left to preserve the state of such historical works.

Forgotten Books is a registered trademark of FB &c Ltd.
Copyright © 2018 FB &c Ltd.
FB &c Ltd, Dalton House, 60 Windsor Avenue, London, SW19 2RR.
Company number 08720141. Registered in England and Wales.

For support please visit www.forgottenbooks.com

1 MONTH OF FREE READING

at
www.ForgottenBooks.com

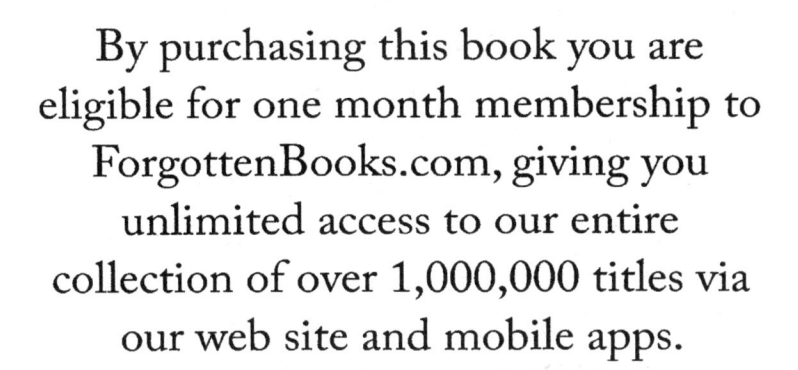

By purchasing this book you are eligible for one month membership to ForgottenBooks.com, giving you unlimited access to our entire collection of over 1,000,000 titles via our web site and mobile apps.

To claim your free month visit:
www.forgottenbooks.com/free857397

* Offer is valid for 45 days from date of purchase. Terms and conditions apply.

English
Français
Deutsche
Italiano
Español
Português

www.forgottenbooks.com

Mythology Photography **Fiction**
Fishing Christianity **Art** Cooking
Essays Buddhism Freemasonry
Medicine **Biology** Music **Ancient Egypt** Evolution Carpentry Physics
Dance Geology **Mathematics** Fitness
Shakespeare **Folklore** Yoga Marketing
Confidence Immortality Biographies
Poetry **Psychology** Witchcraft
Electronics Chemistry History **Law**
Accounting **Philosophy** Anthropology
Alchemy Drama Quantum Mechanics
Atheism Sexual Health **Ancient History**
Entrepreneurship Languages Sport
Paleontology Needlework Islam
Metaphysics Investment Archaeology
Parenting Statistics Criminology
Motivational

[*All Rights Reserved.*

MANUAL

OF

FIELD SKETCHING

AND

RECONNAISSANCE.

LONDON:
PRINTED FOR HIS MAJESTY'S STATIONERY OFFICE,
BY HARRISON AND SONS, ST. MARTIN'S LANE,
PRINTERS IN ORDINARY TO HIS MAJESTY.

And to be purchased, either directly or through any Bookseller, from
EYRE & SPOTTISWOODE, EAST HARDING STREET, FLEET STREET, E.C.;
or OLIVER & BOYD, EDINBURGH;
or E. PONSONBY, 116, GRAFTON STREET, DUBLIN.

Price One Shilling and Sixpence.

UG
470
G78
1903

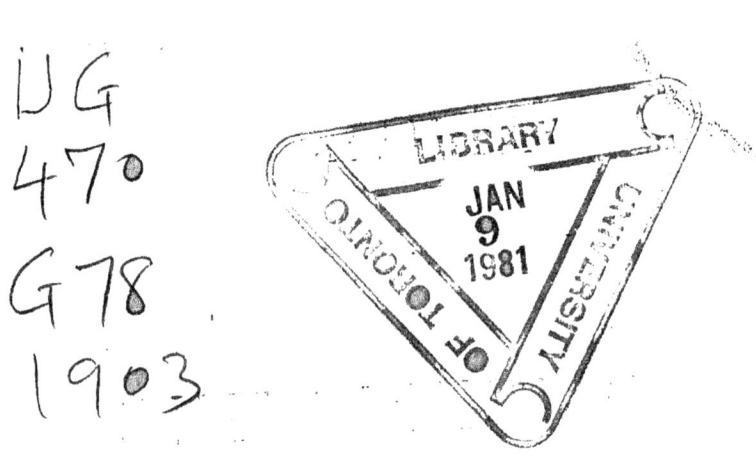

(Wt. 23446 3000 1|04—H & S 3845) P. 02/847

RECONNAISSANCE.

CORRIGENDA.

The following corrections and additions should be made in the Manual :—

Title page—Add the following note under title of Book, "This Manual supersedes the Text Book of Military Topography, Part I."

Page 23—The last paragraph should read : " Every main road or railway should have From ——— printed at the left end of it, on the margin of the sketch, and To ——— at the right end. The distance between the nearest town or village and the margin of the sketch is sometimes given thus—From SELLING 2 miles, To CANTERBURY 3½ miles."

Page 26—In the last two *lines* the names COBHAM, CUXTON, CANTERBURY should be in block type capitals, COBHAM. *Vide* note on *page* 92.

Page 38, *line* 13, read—" Text book of Topographical Surveying, 1904."

Page 47, *line* 3—For " ars " read " are."

Line 8—For " positiod " read " position."

Line 10—Insert hyphens between 1-2, 1-3, 1-4 and 1-5.

Plate XI, facing *page* 49—Erase the heading, "Text Book of Military Topography, Part I."

Page 52, *lines* 26 and 27 should read—" The declination is subject to two principal variations, annual and positional."

Page 52—Omit last two *lines*.

Page 53—Omit first five *lines*.

Page 54, *line* 12, the reference should read—" Text Book of Topographical Surveying, 1904."

Lines 17 to 20—Omit words from " But " to " date " and add " If the magnetic declination is known, the true north line should be shown in addition to the magnetic north line ; but the magnetic declination should not be determined by any approximate methods."

Lines 34 and 35—Omit the words from " or where " to " planetable."

Page 88, *line* 29—For " V and VI " substitute " v and vi."

Page 92—The names LONDON, WELLINGTON, &c., should be in ordinary block capitals, LONDON.

Plate XIX, facing *page* 99—Alter misprint in spelling of word " Reconnaissance " in heading of *Plate* XIX.

Page 116—Last *line* read " Text Book of Topographical Surveying, 1904."

CONTENTS.

			PAGE
Chapter	I.	Introduction.—The uses of Military Maps and Sketches	5
,,	II.	Definitions.—1. Topographical Forms; 2. Technical Terms	8
,,	III.	Scales	12
,,	IV.	Copying, Reducing and Enlarging Maps..	19
,,	V.	Field Sketching. Conventional Signs. Lettering	23
,,	VI.	,, ,, The Representation of Hill Features	27
,,	VII.	,, ,, Instruments used	35
		The Plane-table..	38
,,	VIII.	,, ,, The Prismatic Compass	50
,,	IX.	,, ,, Traversing and the Field Book	63
,,	X.	,, ,, Sketching on Horseback	70
,,	XI.	,, ,, With the aid of Range-finders	73
,,	XII.	,, ,, Landscape Sketching. Eye-sketching	79
,,	XIII.	Night Marches. Map Reading	83
,,	XIV.	Reconnaissance	90

APPENDIX.—Approximate Methods of finding the True North 116
INDEX.. 117

PLATES.

			PAGE
Plate	I.	Scales	15
,,	II.	Scales	16
,,	III.	Conventional Signs used on Ordnance Maps	23
	IV.	,, ,, ,, in Field Sketching	24
	V.	,, ,, Lettering	24
	VI.	,, ,, Hill Drawing	26
,,	VII.	Photograph of Model of Country	27
,,	VIII.	Ordnance Map, 1 inch to 1 mile, coloured	28
,,	IX.	Service Plane-tables	38
,,	X.	Small Scale work with Form-lines	49
	XI.	Hill Drawing, showing importance of marking Watercourses	49
,,	XII.	Field Sketch, 1 inch to 1 mile, pencil	49
,,	XIII.	Diagram of Magnetic Variation	52
,,	XIV.	Specimen Page of Compass Traverse	66
,,	XV.	Plotting of above Trasverse	66
,,	XVI.	Plane-table and Range-finder Sketch	78
,,	XVII.	Landscape Sketch made in Afghan War, 1880	79
,,	XVIII.	,, ,, ,, S. African War, 1899	80
,,	XIX.	Road Sketch and Report	99
,,	XX.	Dervish Map of Fort at Kosheh	116

MANUAL
OF
FIELD SKETCHING
AND
RECONNAISSANCE.

CHAPTER I.

INTRODUCTION.

AN officer should commence the study of field sketching with a clear idea of the purposes for which military maps and sketches are used.

The *first* and most obvious use of a map or sketch is to enable the user to find his way in a country unknown to him; this is the use to which travellers of all sorts put maps, and maps are a vital necessity to an army on the march.

The *second*, is to give the officer in command of a moving body of troops an idea of the essential features of the country through which he is marching and of which he can only see a small portion, what hills and rivers must be crossed, what are the alternative routes, where can water, fuel, shelter, be got, what defiles must be traversed, and the thousand questions arising on the march which can be answered by reference to a map.

The *third*, is to enable a general in command of extended operations to obtain a diagrammatic or bird's-eye view of the scene of them on which to base his strategy.

The *fourth*, is to assist in the execution of tactical dispositions and operations in laying out, for instance, a general defensive or outpost position; or to enable the officer commanding one portion of, or unit in, a line of defence, to see the disposition of the remainder.

The *fifth*, is to illustrate a report, such as a reconnaissance or outpost report, which would be hardly intelligible without an accompanying sketch.

There are other uses to which military maps and sketches may be put, but one general principle covers all, and this is, that maps and sketches are required in war to enable officers to ascertain the natural and artificial features of country which is either not within their sight or is too far off for the features to be clearly recognisable, or which is of too complicated a nature to be understood without the aid of a map; in fact, to give information which cannot be obtained at the moment by the eye alone.

There are certain uses to which maps should not be put. Maps should never be used for the actual selection of the exact position of trenches or other field defences or of outpost sentries; these should always be determined on the ground. A military map or sketch, however, is invaluable for the selection of a general position of defence, attack or observation.

Except in the case of absolute savages no enemy will be met who does not use maps and sketches. Interesting examples of maps used by the enemy are, the Boer maps of the Ladysmith and Tugela country, made during the siege; the map of Sikkim, made by the natives in the campaign of 1889; and the Dervish map of the fort at Kosheh, made during the campaign of 1885.

A **map** is the representation on paper of a portion of the earth's surface.

A **topographical map** shows artificial features such as roads, railways and villages, and natural features such as hills, valleys, plains, rivers, streams and lakes.

Field sketching is the art of producing a topographical map suitable for military purposes.

A **military sketch** is a rough topographical map on which only features of military importance are represented.

The *information* given upon maps is limited by their scale. For instance, detail that can be shown upon a map on a scale of two inches to a mile, cannot be represented on a map on a scale of one inch to a mile. The reliance that can be placed on a map depends upon the nature and date of the survey or last revision.

No false information, nothing but what has been seen should be represented, and the tendency to sacrifice truth to effect must be guarded against.

An important habit engendered by military sketching is that of studying the forms of ground and thinking of their effect on military operations. *Map reading*, or the faculty, which every officer should possess, of quickly grasping the correct meaning of a map or military sketch, cannot be acquired without study. An officer who can sketch is most likely to appreciate and make good use both of existing maps and of the sketches produced by others.

The first requirement that a military sketch should fulfil is *intelligibility*. Minute accuracy of position is of less importance, although all attainable accuracy that the circumstances allow should be striven for. Intelligibility includes clearness and simplicity, and and simplicity consists in the omission of unnecessary detail.*

Maps consist of *outline*, that which can be shown in plan; of *detail*, such as woods, marshes, rocks, and hill-features, which are shown conventionally; and of *writing* or lettering, names, remarks and figures.

On all maps and sketches, the features of the ground and any detail that, from want of space, cannot be shown in plan, are represented conventionally. Thus roads, rivers, lakes, buildings, &c., are shown by lines representing their outline or plan; the irregularities of the ground are conventionally represented by contours, form-lines, shading or hachures, and woods, marshes, railways, bridges, &c., are distinguished by *conventional signs*.

---

* If, from any point on the ground represented on a military sketch, an exceptionally good general view can be obtained, a note to that effect should be made on the sketch.

CHAPTER II.

DEFINITIONS.

1. Topographical Forms.

Basin A term used to describe (*a*) a small area of level ground surrounded or nearly surrounded by hills; and (*b*) a district drained by a river and its tributaries, as the "basin of the Thames."

Crest: The edge of the top of a hill or mountain; the position at which a gentle slope changes to an abrupt one; the top of a bluff or cliff.

Col: A depression between two adjacent mountains or hills; or a break in a ridge; or the neck of land which connects an outlying feature with a range of mountains or hills, or with a spur. In most foreign maps this feature is represented by a special conventional sign, an arrangement of dotted form-lines, thus :—

FIG. 1.

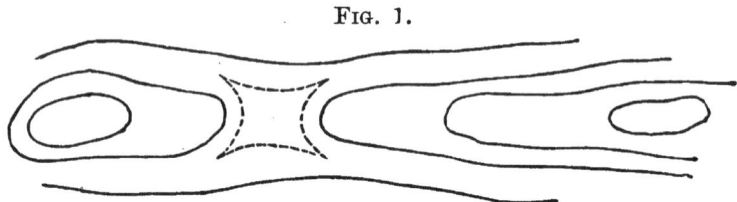

Dune: A hill or ridge of sand formed by the winds near a sea or lake shore, along a river bed or on a sandy plain.

Escarpment: An extended line of cliffs or bluffs.

Gorge: A rugged and deep ravine.

Knoll: A low hill.

Pass: A depression in a mountain range through which a road or trail may pass; a track over a mountain range.

Plateau: An elevated plain.

Saddle: A col.

Spur: A projection from the side of a hill or mountain running out from the main feature.

Thalweg: A watercourse; a valley bottom; the deepest line of a valley along which water would flow.

Watercourse: The line defining the lowest part of a valley, whether occupied by a stream or not.

Watershed: A ridge of high land separating two drainage basins; the summit of land from which water divides or flows in two directions. A watershed does not necessarily include the *highest points* of a chain of mountains or range of hills.

This list does not profess to be exhaustive; there are many terms in common use such as hill, mountain, river, slope, island, cliff, underfeature, &c., which it does not appear necessary to define.

2. *Technical Terms.*

Angle: Back-angle, the direction or bearing in a traverse of a station which has been passed.

Closing-angle, the angle taken from the last station in a traverse to some fixed point, to ascertain whether the traverse closes satisfactorily.

Forward-angle, the forward direction or bearing of one station in a traverse to the next in succession.

Base, or **Base-line:** A carefully chosen and accurately measured line upon which the accuracy of a triangulation depends.

Bearing: True bearing is the angle a line makes with the true north line.

Magnetic bearing is the angle a line makes with the magnetic north line.

Contour: A line joining all points on a map representing points on the earth's surface which are the same height above mean-sea-level. Or thus:—A contour is an imaginary line running along the surface of the ground at the same level throughout its length. Contours may also be defined as the plans of the lines at which a water surface (of the ocean, for instance) would intersect the surface of the earth were it raised successively by equal amounts.

Datum or **Datum-level:** An assumed level to which altitudes are referred.

Fall of a river: Its slope, usually measured in inches (or feet) per mile; thus 9 inches per mile.

Field-book: Any book in which a surveyor enters observations or measurements in the field. Usually applied to a book used in traversing.

Form-line: An approximate contour; a sketch contour.

Gradient: A slope expressed by a fraction. Thus $\frac{1}{30}$ represents a rise or fall of 1 foot in 30 feet.

Hachures: Vertical hachuring is a conventional method of representing hill features by shading in short disconnected lines drawn directly up and down the slopes in the direction of the flow of water on the slopes.

Horizontal Equivalent: Sometimes written H.E., is the distance in plan between two adjacent contours measured in yards.

Local Magnetic Attraction: Is the deviation of the magnetic needle of a compass from its mean position owing to the presence of masses of magnetic iron-ore or of iron in the neighbourhood.

Magnetic Declination: (In military surveys commonly called magnetic variation.) The angle between the true and the magnetic meridians. This is called E. or W. declination, according as the magnetic N. is E. or W. of the true N. The variation is strictly the change in the value of the declination.

Meridian or **Meridian-line**: A true north and south line.

Magnetic Meridian: A magnetic north and south line.

Offset: A short measurement at right angles to a main measured line.

Orienting or **Setting** a map or plane-table is the process of placing the map or plane-table so that the north line points north.

Plotting: The process of laying down on paper field observations and measurements.

Resection: A method by which the sketcher determines his position by observing the bearings of, or drawing lines from, at least two previously fixed points.

Section: The outline of the intersection of the surface of the ground by a vertical plane.

Traverse: The survey of a road, river, or track by measuring a continuous series of straight lines along its course and the angles at their junctions.

Triangle of Error: The small triangle formed by the three rays when resecting on a plane-table which is not properly oriented.

Triangulation: The process of fixing the position of points on the area to be surveyed, by means of a measured base and a chain or network of triangles dependent on it.

Vertical Interval: Sometimes written V.I., and always given in feet, is the difference of level between two adjacent contours.

CHAPTER III.

SCALES.

THE word **scale** is used to denote the proportion which a distance between any two points on a sketch or map bears to the horizontal distance between the same two points on the ground. Thus, if the distance between two farms on a map be 1 inch, and the horizontal distance on the ground be 2 miles, the scale of the map will be 1 inch to 2 miles. Similarly the statement that a map is on a scale of $\frac{1}{250000}$ implies that a distance of 1 inch on the map represents an actual distance of 250,000 inches, or 3·94 miles.

The increased range of modern weapons has affected the question of the scales of military maps and sketches. It is obvious that if troops come into action at longer ranges than formerly, sketches must show larger areas, and the scales must consequently be smaller. Again, in making a reconnaissance it is no longer possible to approach the enemy so closely as in times past. To give an example:—In the early part of the 19th century, when it was necessary to make a reconnaissance, "the officers obtained half a squadron of troops, dressed in the gayest possible uniform, which cantered to within 200 or 300 yards of the enemy's positions and made careful observations, knowing that they were practically quite safe."

Again, scales which were suitable for use in the days of concentrated fights like Waterloo would, if employed now, show but a small portion of the battle-field.

Then, outposts are further from the main body than in old days. Forts and defences are more widely spaced. A fortress in the middle of the 19th century might be two or three miles in circumference; nowadays these figures must be multiplied by 10 and the enclosed area by 100.

Troops are more widely spaced both in attack and defence. It is therefore clear that the scales of military maps and sketches must be smaller than in the past.

Nowadays, although it would be going too far to say that large scales will never be used, their importance has diminished. In the South African War there was hardly an instance of the use of the 6-inch scale, nor in recent Indian campaigns, nor in China. Whilst the value of large scale work has decreased, the importance of the smaller scales has greatly increased. It should be borne in mind that a general does not as a rule require a sketch for ground of which the details are clearly visible to him. The tactical use of a sketch is specially to give information about ground too distant to see or which is hidden by intervening hills or woods. A grave objection to the use of large scales is the actual mechanical difficulty of carrying the maps when made. (This does not apply to siege or similar operations.)

It is of course clear that in "open" country, i.e., country free from much detail, the scales should generally be smaller than in "close" country. But even in England, which is perhaps the closest country in the world, it is found that a scale of 1 inch to 1 mile is sufficiently large for general use, and a scale of $\frac{1}{2}$ inch to 1 mile has been adopted for manœuvres, whilst a scale of 2 inches to 1 mile may be taken as the most useful scale for local maps and sketches. In India the frontier maps are mostly on a scale of $\frac{1}{4}$ inch to 1 mile, though some are on as small a scale as $\frac{1}{8}$ inch to 1 mile, and in countries where the hill features are large and the artificial features few, these scales are perhaps large enough.

It is not necessary to lay down a very rigid rule as to the scales on which military sketches should be made, but the following will serve as a guide :—For *local maps*, generally for defensive or outpost positions, 2 or 4 inches to 1 mile; for a general map or sketch of a *district* from 1 inch to $\frac{1}{4}$ inch to 1 mile; for a *road* or *river* sketch from 1 to 2 inches to a mile.

Special sketches such as those which may be required for the defence of *villages* or *towns*, or for selection of camps, or for a section of the deliberate investment of a *fortress*, may be on large scales of from 4 inches to a mile and upward; these, in the case of a siege, are more likely to be carried out by expert topographers.

"Large scales, such as 6 inches to the mile, should be sparingly used for instruction, and never for tactical exercises."[*]

---

[*] Regulations as to the Issue of Military Maps. 1902.

At home and in India and the Colonies the scale of a map is usually expressed in words, showing the relation between inches on the map and miles on the ground, thus: 1 inch to 1 mile, ¼ inch to 1 mile, &c. The advantage of this method is that the eye, having once been trained to recognize the length of an inch on paper, can readily estimate the distance between any two points on a map with considerable accuracy.

Foreign maps are usually constructed on scales which bear the proportion of 1 to some multiple of 10, such as 1 to 100,000, 1 to 500,000, written thus: 1 : 100,000, or 1 : 500,000. This proportion is shown on the map by a **Representative Fraction** (R.F.), as $\frac{1}{100000}$ or $\frac{1}{500000}$, which means that the distance between any two points in the map is $\frac{1}{100000}$th or $\frac{1}{500000}$th of the distance between the same points on the ground. The numerator of the R.F. is always 1. If the R.F. of a map be $\frac{1}{63360}$, 1 inch on the map will be equal to 63,360 inches, or 1 mile on the ground.

Example.—If the scale of a map is 2 inches to a mile:

$$\text{R.F.} = \frac{2 \text{ inches}}{1 \text{ mile}} = \frac{2 \text{ inches}}{63{,}360 \text{ inches}} = \frac{1}{31{,}680}.$$

On all British maps the scale is now shown by a straight line divided into several equal parts; by the R.F.; and by a verbal description.

The measure of length which a scale is to show, whether feet, yards or miles, is termed the *unit of measure*, and scales are usually, though not necessarily, constructed of such a length as to represent a distance which is a multiple of ten such units, as 100 feet, 50 yards, 80 miles.

A scale should usually be from 4 to 6 inches long, and to construct it the number of tens, hundreds, or thousands of the "unit of measure", which will occupy such a length on paper, should be calculated. Such a scale is called a *Plain scale*.

To face page 15.

Plate I.

Scale $\frac{1}{10560}$, or 6 inches to 1 mile

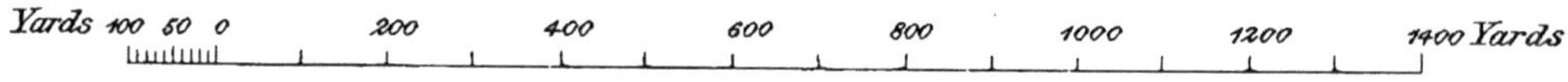

Yards 100 50 0   200   400   600   800   1000   1200   1400 Yards

Scale $\frac{1}{31680}$, or 2 inches to 1 mile

Yards 1000   500   0   1000   2000   3000   4000

Scale $\frac{1}{100000}$, or 1·58 miles to 1 inch.

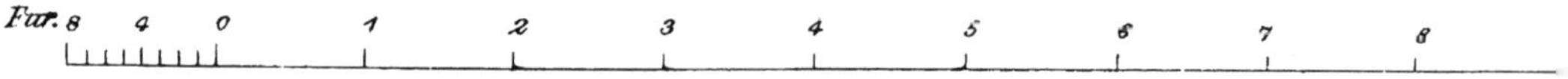

Fur. 8   4   0   1   2   3   4   5   6   7   8

EXAMPLES.

Ex. 1.—**To construct a scale of 6 inches to 1 mile to show tens of yards.**

Here 6 inches represents 1 mile, or 63,360 inches:—

$$\text{The R.F.} = \frac{6}{63,360} \text{ or } \frac{1}{10,560}.$$

If the whole length of the scale is to represent 1,500 yards, the length of this on paper will be $\frac{1500}{10,560}$ yards, or $\frac{1500 \times 36}{10,560}$ inches, or 5·11 inches,

or, if it is preferred, the result may be obtained from a proportion sum, thus:—

Yards in a mile.	Yards in the scale.		Inches representing a mile.		Length of scale in inches.
1,760 :	1,500	: :	6	:	5·11

In calculating the length of a scale it should be worked out to two places of decimals of an inch.

To draw the scale.—Draw a line 5·11 inches long, and divide it into 15 equal parts, each representing 100 yards. These parts are called the *primary divisions* of the scale. Subdivide the left division into 10 parts, each representing 10 yards, called *secondary divisions*. The zero of the scale should be placed as shown (Plate I), and the " primary " and " secondary " divisions numbered outward from it.

A diagonal scale of inches will be found on most protractors from which the required lengths can be measured. Or take a pair of dividers and divide an inch into tenths and twentieths by trial and error, the smaller fractions may be estimated. An explanation of the construction and use of diagonal scales is given on p. 17.

Ex. 2.—**To construct a scale of 2 inches to 1 mile to show hundreds of yards.**

It is convenient to make scales about 4 to 6 inches long; in this case 5,000 yards will clearly be represented by a length of nearly 6 inches and this will do.

$$\text{The R.F.} = \frac{2 \text{ inches}}{1 \text{ mile}} = \frac{2 \text{ inches}}{1{,}760 \times 36 \text{ inches}} = \frac{1}{31{,}680}.$$

Then 5,000 yards will be represented by $\dfrac{5{,}000 \times 36}{31{,}680}$ inches,

or 5·68 inches.

Draw a line 5·68 inches long.

Divide this into 5 equal parts (each being therefore 1·14 inches long).

Subdivide the left division into 10 equal parts, each 0·11 inches long; each of these will represent one hundred yards. (Plate I.)

Ex. 3.—**To construct a scale of 1 inch to 1 mile to show miles and furlongs.**

Here it is clear that all that has to be done is to take a length of say 6 inches representing 6 miles, divide this into inches and the left division into eighths.

Ex. 4 —**To construct a scale of $\frac{1}{100000}$ to show miles.**

Here 1 mile is represented by $\frac{1}{100000}$ mile.

i.e., by $\dfrac{5{,}280 \times 12}{100{,}000}$ inch,

or 0·634 inch.

Hence, if the whole scale represents 10 miles, it will be 6·34 inches long, each mile will be 0·63 inches long, and if the left division is divided into eighths to show furlongs, each of these small divisions will be 0·08 inch long. (Plate I.)

Comparative Scales express the same proportion in different units of measure, or in other words, are scales having the same R.F. but a different graduation. A simple instance will show their application:—

A Frenchman and an Englishman sketch the same portion of Russian ground to a scale of $\frac{1}{10000}$. The sketches when finished will be of exactly the same size, but the French sketch will show a to the ninth above, and so on, till we come to a line

$$\text{The R.F.} = \frac{2 \text{ inches}}{1 \text{ mile}} = \frac{2 \text{ inches}}{1{,}760 \times 36 \text{ inches}} = \frac{1}{31{,}680}.$$

Then 5,000 yards will be represented by $\dfrac{5{,}000 \times 36}{31{,}680}$ inches,

or 5·68 inches.

Draw a line 5·68 inches long.

Divide this into 5 equal parts (each being therefore 1·14 inches long).

Subdivide the left division into 10 equal parts, each 0·11 inches long; each of these will represent one hundred yards. (Plate I.)

Ex. 3.—**To construct a scale of 1 inch to 1 mile to show miles and furlongs.**

Here it is clear that all that has to be done is to take a length of say 6 inches representing 6 miles, divide this into inches and the left division into eighths.

Ex. 4 —**To construct a scale of $\frac{1}{100000}$ to show miles.**

Here 1 mile is represented by $\frac{1}{100000}$ mile.

i.e., by $\dfrac{5{,}280 \times 12}{100{,}000}$ inch,

or 0·634 inch.

Hence, if the whole scale represents 10 miles, it will be 6·34 inches long, each mile will be 0·63 inches long, and if the left division is divided into eighths to show furlongs, each of these small divisions will be 0·08 inch long. (Plate I.)

Comparative Scales express the same proportion in different units of measure, or in other words, are scales having the same R.F. but a different graduation. A simple instance will show their application:—

A Frenchman and an Englishman sketch the same portion of Russian ground to a scale of $\frac{1}{10000}$. The sketches when finished will be of exactly the same size, but the French sketch will show a

To face page 16.

Plate II.

Scale $\frac{1}{10000}$ or 6·34 inches to 1 Mile.

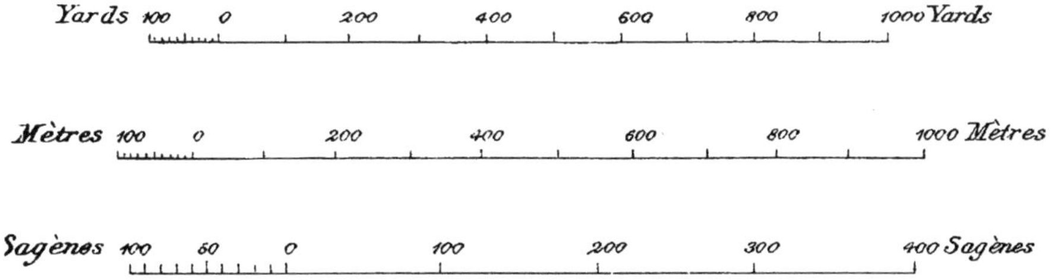

Yards 100 0 200 400 600 800 1000 Yards

Mètres 100 0 200 400 600 800 1000 Mètres

Sagènes 100 50 0 100 200 300 400 Sagènes

scale of metres. The R.F. is common to both, and the scales are, therefore, comparative. Should the sketches now fall into the hands of a Russian, who was ignorant of the units of length marked on their scales, he would be unable to utilize them for measurement until he had constructed from the R.F. another comparative scale showing Russian units of length. See Plate II.

In military operations distances are occasionally measured by the time required to traverse them, and in this case the linear scale may be usefully supplemented by a comparative scale of hours.

Ex. 5.—**To construct a scale of time for a column of troops marching at the rate of 3 miles an hour.**

As an example, the scale may be taken as 1 inch to 3 miles, which gives

$$\text{R.F.} = \frac{1}{63{,}360 \times 3} = \frac{1}{190{,}080},$$

1 inch will represent 1 hour, and by dividing the inch into 12 parts 5 minutes may be shown.

Diagonal Scales are used when it is necessary to measure smaller dimensions than those into which the left-hand primary division of a scale can be conveniently divided.

The principle governing the construction of these scales is that "in similar triangles the sides are proportional."

It is plain that a line 1 inch long cannot be divided directly into 100 equal parts, but it can be divided into 10 equal parts, and by the diagonal method a tenth of each of these parts, or the $\frac{1}{100}$th part of an inch is obtained.

Draw 11 parallel equidistant lines (Fig. 2); divide the lower of these lines into equal parts of the intended length of the primary divisions, say, 1 inch; at each of these divisions erect a perpendicular, cutting all the 11 parallels; and number these primary divisions, 1, 2, 3, beginning from the second.

Divide the first primary division AB into 10 equal parts, both upon the highest and lowest of the 11 parallel lines, and let these subdivisions be reckoned in the opposite direction to the primary divisions, as in the simply-divided scales. Draw the diagonal lines from the ninth subdivision below to the tenth above; from the eighth below to the ninth above, and so on, till we come to a line from the zero point below to the first subdivision above. The scale is now complete, and the smallest division on it is EF, which is

FIG. 2.

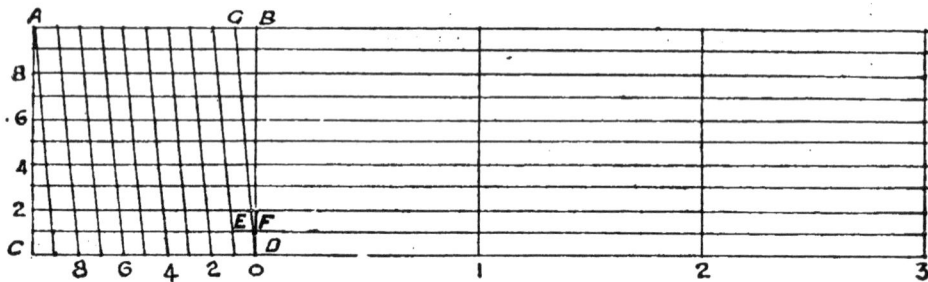

$\frac{1}{100}$th of an inch. GBD and EFD are similar triangles, therefore their sides are proportional and consequently

$$BD : FD :: GB : EF$$
$$1 : \tfrac{1}{10} :: \tfrac{1}{10} : \tfrac{1}{100}$$

Diagonal scales of *half an inch*, and *a quarter of an inch*, constructed as above will give readings of $\frac{1}{200}$, or ·005 of an inch, and $\frac{1}{400}$, or ·0025 of an inch respectively. The latter scale is mentioned as it is given on many protractors, but the divisions are so minute that it is seldom used for military purposes.

The **sketching protractor** usually has the scales of 6 and 8 inches to a mile engraved on it, and it is equally suitable for sketching on scales which are multiples and measures of these numbers; *e.g.*: For sketching at 3 inches to a mile, divide the measured distance by two, and take it off the scale of 6 inches to a mile—or for 2 inches to a mile, divide the distance by four and use the scale of 8 inches to a mile.

Measurements in the field should be taken direct from the protractor without using compasses.

CHAPTER IV.

Copying, Reducing, and Enlarging Maps.

Copying.—Maps which are to be simply copied without reduction or enlargement may be first traced on tracing paper, and then a piece of carbon paper being placed between the tracing and a clean sheet of paper, and both very firmly fastened down by drawing pins, the tracing may be transferred by following the lines with a pointed instrument.

In the field an excellent substitute for carbon paper can be made by rubbing lead pencil dust on a piece of tracing paper; in fact this is to be preferred to carbon paper. Any sharp piece of wood will do for the tracing instrument.

A map may also be transferred direct without a tracing, but there is some danger of damaging the original by the point.

Tracing cloth is useful for copying; it is tough and will bear much handling. Draw on the glazed side, and put washes of water colour, if required, on the back, darker than they are intended to appear on the front.

Reducing or Enlarging.—It sometimes happens that a reduction or enlargement is required of some road or river sketch, a track of some length and of no great breadth. The following is in this case a more rapid and accurate way of doing what is required than by means of squares. The method can best be explained by means of an example (see Fig. 3).

Suppose it is required to reduce a map of a river ACB to $\frac{1}{3}$ of its linear scale. Take any point O well outside of AB; join OA, OB; join important points on ACB (such as the bend C) with O.

Make $Oa = \frac{1}{3}$ OA and $Ob = \frac{1}{3}$ OB; through a draw ac parallel to AC; the intersection of ac and OC gives the required point c and so on for the other points.

If a proportional compass is available, by its means plans may be enlarged or reduced so that all lines on the new map shall bear any required ratio to lines on the old map. But in the field these instruments will not usually be found when wanted, and the quickest way is to draw squares of any convenient size on the original, and then rule the paper on which the new map is to be

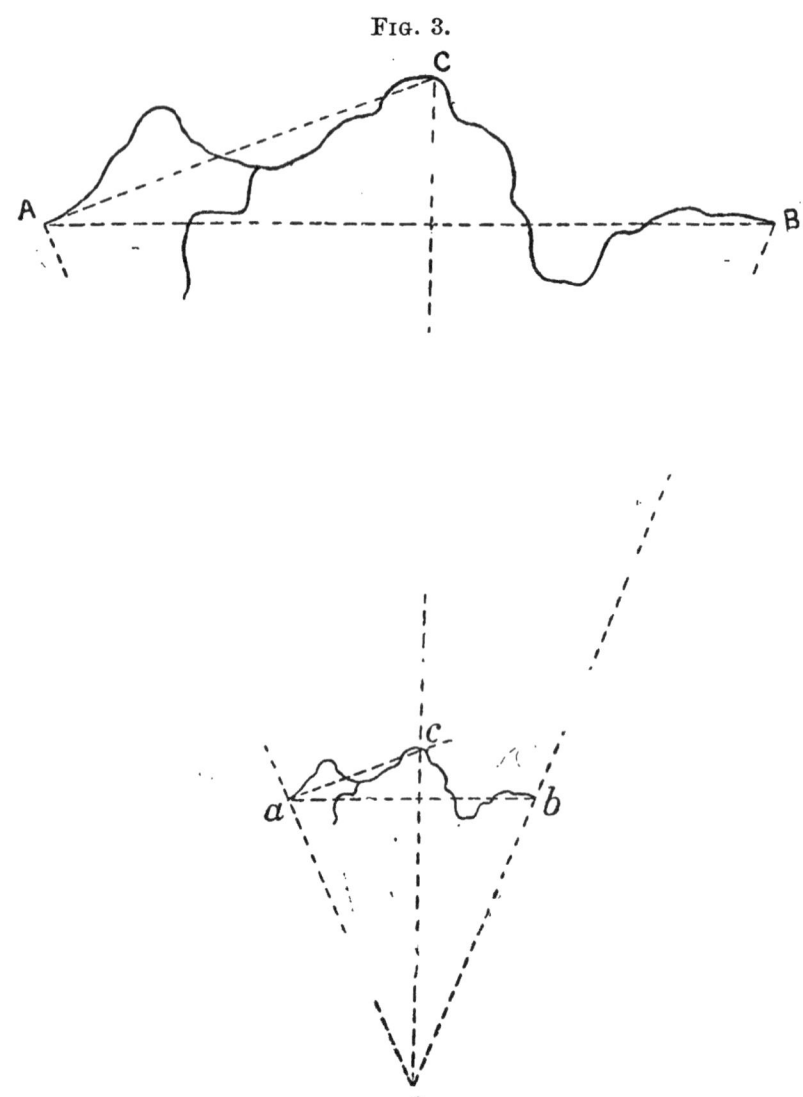

FIG. 3.

made with squares whose sides bear the required ratio to the sides of the squares on the original. Then copy the original by eye, so that objects occupy the same positions relatively to the squares as they did in the original.

FIG. 4.

A

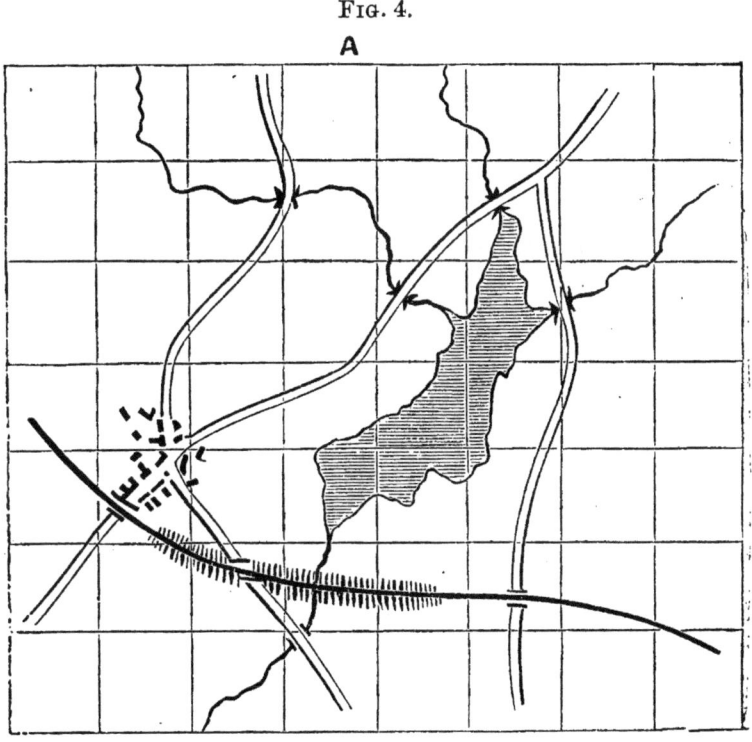

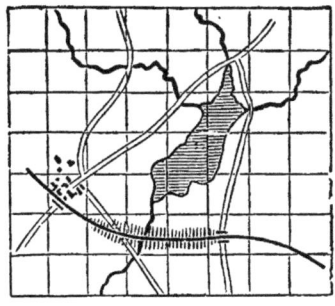

Ex. 1.—It is required to reproduce at 2 inches to a mile a plan **A** (Fig. 4), the scale of which is 4½ inches to a mile. The length of the plan A is 3·25 inches; thus—

$$4\cdot 5 : 2 :: 3\cdot 25 : 1\cdot 44 \text{ inches.}$$

Divide the length of each of them into the same number of equal parts, say 8. Raise perpendiculars. On these set off as many of the same equal parts as may be required, and join the points so as to form the squares.

Ex. 2.—A plan with the R.F. $\frac{1}{25000}$ is to be copied at 6 inches to a mile. The plan is 10 inches long. The R.F. of 6 inches being $\frac{1}{10560}$, the length of the copy will be—

$$\frac{1}{25000} : \frac{1}{10560} :: 10 : 23\cdot 67 \text{ inches.}$$

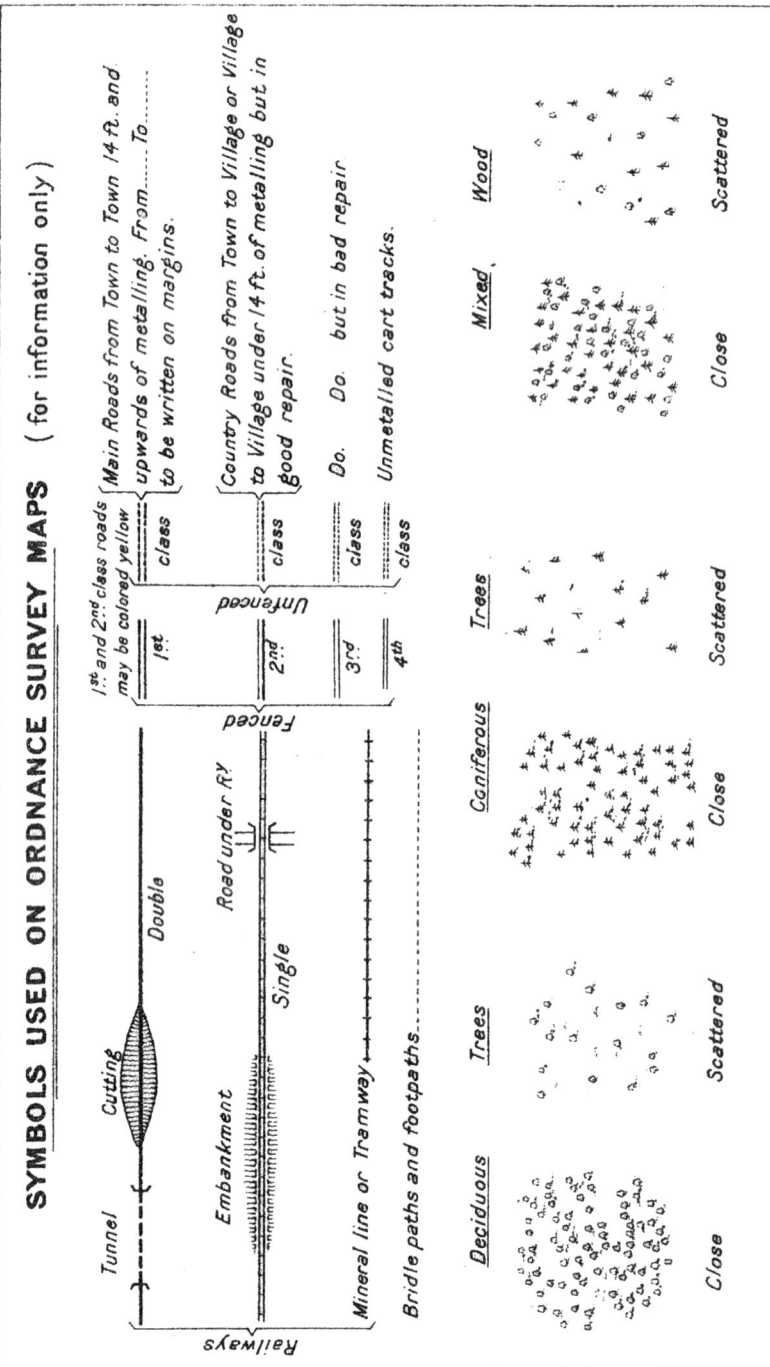

to face p. 23
PLATE III
CONVENTIONAL SIGNS PLATE Nº I.

SYMBOLS USED ON ORDNANCE SURVEY MAPS (for information only)

Railways
- Tunnel
- Cutting
- Double
- Embankment
- Road under R?
- Single

Mineral line or Tramway

Bridle paths and footpaths

Fenced
- 1st class — Main Roads from Town to Town 14 ft. and upwards of metalling. From .... To .... to be written on margins.
- 2nd class — Country Roads from Town to Village or Village to Village under 14 ft. of metalling but in good repair.
- 3rd class — Do. Do. but in bad repair.
- 4th class — Unmetalled cart tracks.

Unfenced
- 1st and 2nd class roads may be colored yellow

Deciduous Trees — Close, Scattered

Coniferous Trees — Close, Scattered

Mixed Wood — Close, Scattered

Post & Telegraph Office, T.   Post Office, P.

CHAPTER V.

FIELD SKETCHING.

Conventional Signs and Lettering.

Conventional Signs.

Conventional signs enable the draughtsman to give an amount of information on a sketch, or map, that could not, conveniently, be otherwise conveyed. They should be simple in character, so that they may be easily understood, and not numerous. It is far better to write descriptions on the face of a sketch, in language that cannot be mistaken, than to crowd it with symbols of which the meaning is liable to be misunderstood.

Plates IV, V, VI show the conventional signs approved for military sketches. These signs should be studied, and used in making a finished military sketch. In cases of great haste, however, it is sometimes more convenient to describe the object than to draw its symbol.

The conventional signs used on the one-inch maps of the United Kingdom should also be studied. (Plate III.)

Roads should be drawn with continuous lines when they are fenced, and with dotted lines when they are unfenced.

It should be noted whether a road is metalled or unmetalled, and occasionally it is desirable to give its width. Thus, 15′ m. would mean a metalled road 15 feet wide.

A *railway* is shown by a continuous black line with cross-bars, or by a single red line; the word "single," or "double," should be written along it as the case may be.

Every road or railway should have From —— printed at one end of it, on the margin of the sketch, and To —— at the other end. From —— should be used at the bottom and left of a sketch; To —— should be used at the top and right of a sketch. The distance between the nearest town or village and the margin of the sketch is sometimes given, thus—From Selling 2 miles, To Canterbury 3½ miles.

A *river* should have its name written along its course, and the direction of its current indicated by an arrow. Rivers over 15 feet wide are shown by double lines, under that width by single lines. *Colour* should only be used when a river or stream flows all the year round. The bed of a torrent which flows only in the winter or rainy season should not be shown in colour.

The nature of a *bridge* is indicated by the words iron, stone, wood, suspension, &c.

A *marsh* is shown by lines (blue, when colour is used) parallel to the upper and lower margins of the sketch, and a few touches in black to represent reeds or rushes.

In the case of *woods* it should be noted in writing whether they are passable for troops or not; also their nature, whether of fir, oak, etc.

Postal telegraph offices are distinguished by the letters *P.T.O.*; important *public houses* by *P.H.*; and forges or smithies by *F*.

All *lettering* should be horizontal, excepting the names and direction of rivers, railways, roads, and canals, which should be written along them; also words descriptive of the nature and condition of a tract of country, which should be written so as to, as far as possible, extend over the portion of ground described.

The conventional *colours* are shown on the authorised plate. For rough sketches coloured chalk pencils may be used.

In colouring a finished sketch, the pencil lines are partially rubbed out before the paint is laid on, and all detail is inked in after the colouring is finished.

Troops.—It may occasionally be necessary to show the disposition of troops on a military sketch, as for instance on a sketch of an outpost or defensive position. It is not desirable to spend time in drawing troops in any form to scale. It is sufficient to draw a symbol (filled up with colour, if available), which will attract the attention, writing the necessary information as regards strength and unit alongside.

The following are the recognized symbols for the different arms:—

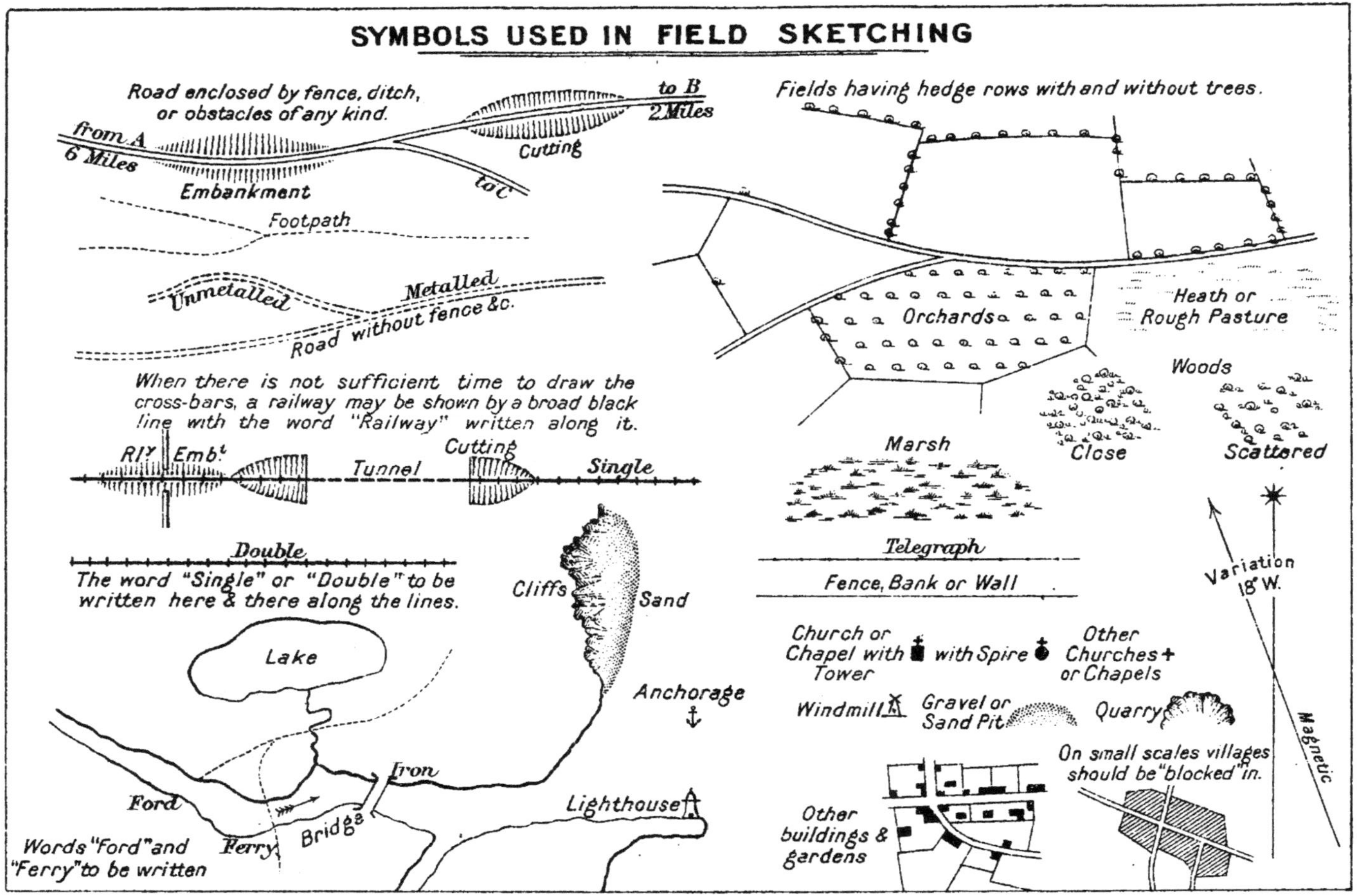

Lithd at the Intell: Div:, War Office, Oct. 1902.

CONVENTIONAL SIGNS PLATE Nº 3.

EXAMPLES OF LETTERING

The "Examples of Lettering" are only "examples" and need not be strictly adhered to, provided always that the lettering be simple and legible.

Scale of Yards $\frac{1}{100,000}$.   Scale of Yards $\frac{1}{100,000}$.

Scale 2 Inches to One Statute Mile.

Cultivation. *Cultivation.* Shingle. *Shingle.* Rough Pasture or Heath.

Bridges. Ferries. Aqueduct. Railway over Road. Embankment. Quarry.

Map of the Neighbourhood of

ABERGAVENNY — SALISBURY

Scale $\frac{1}{31680}$ or 2 Inches to 1 Mile.

Yards 1000   500   0   1000   2000   3000 Yards

Lith^d at the Intell: Div, War Office, Oct. 1902

Cavalry .. ..

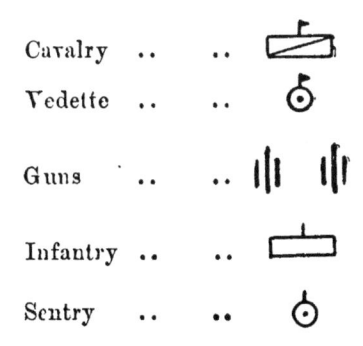

Vedette .. ..

Guns .. ..

Infantry .. ..

Sentry .. ..

The following symbols are also occasionally useful:—

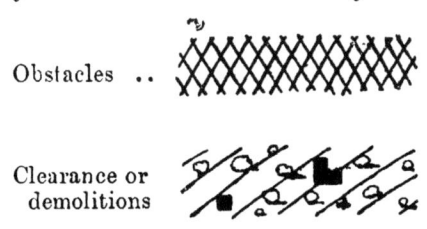

Obstacles ..

Clearance or demolitions

In outpost sketches, the letters P, S, R, may be written instead of piquet, support, and reserve.

The direction of a *patrol* is shown by an arrow.

Intrenchments are shown by double lines on scales of 4 inches to a mile and over, and by a single line on smaller scales. When colour is used these are drawn in red.

Lettering.

It is within the power of all officers to adopt and use some simple style of lettering which shall be clear, and quickly written; italic printing and all fancy strokes should be avoided. Each letter should be separate.

Various specimens are given on Plate V, but officers need not consider themselves bound to use any of the styles there shown, which are only given as illustrations of useful varieties.

Whatever lettering appears on a sketch must be easily legible, and should not interfere with the detail. These are the only essential conditions.

Very bad draftsmen can sometimes usefully employ clear, backward-sloping writing for the minor lettering on a sketch, such as *scale, from, to, single, double, mill, well, broken ground, stone bridge,* &c. The principal words, *i.e.*, the names of towns, villages and rivers may be in simple block character drawn in single strokes. These names may be quite small, provided they are legible, thus:— COBHAM, CUXTON, or larger for more important places, thus:— CANTERBURY. (See Plate V.)

CONVENTIONAL SIGNS PLATE Nº 4.

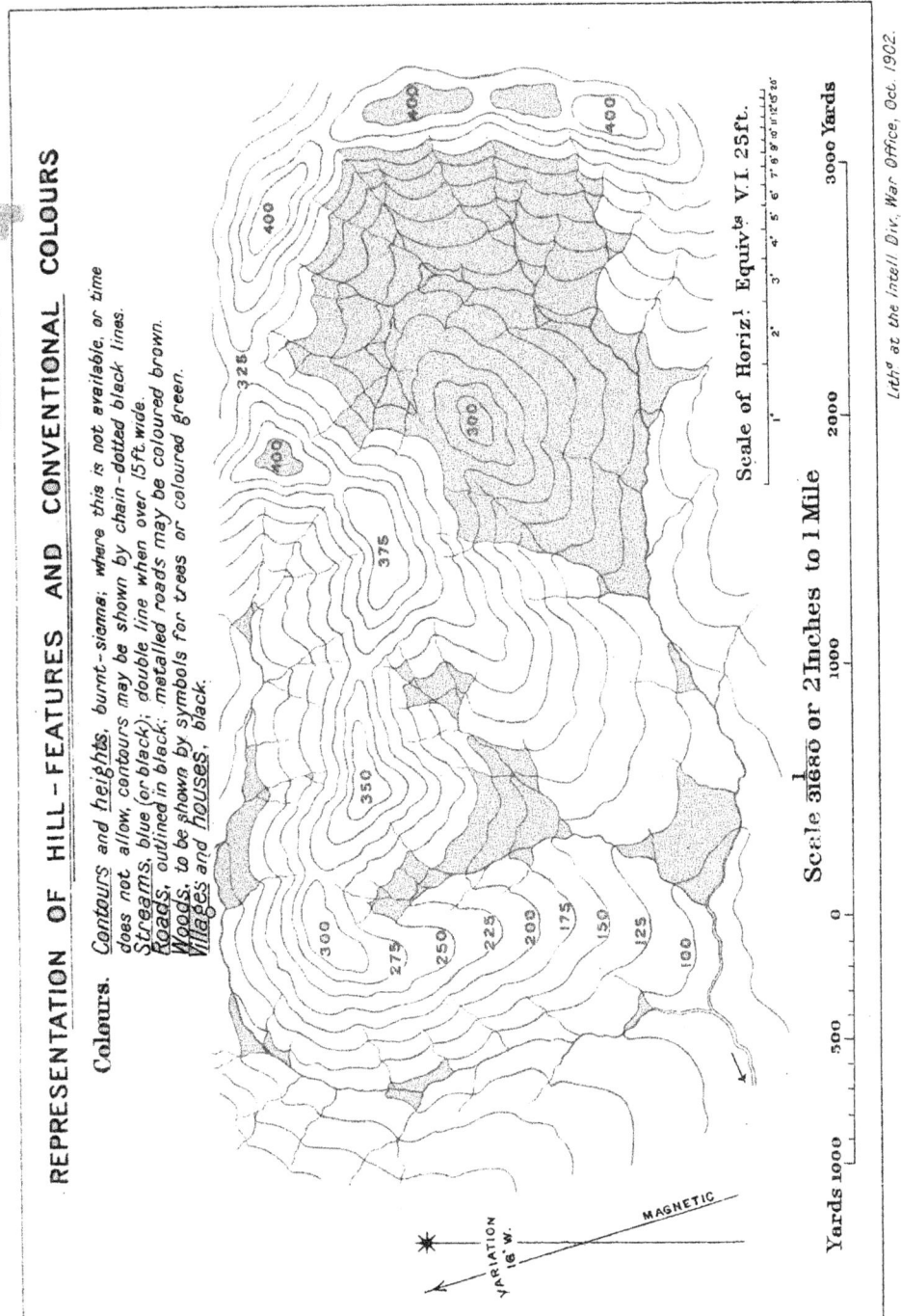

REPRESENTATION OF HILL-FEATURES AND CONVENTIONAL COLOURS

Colours. *Contours* and *heights* burnt-sienna; where this is not available, or time does not allow, contours may be shown by chain-dotted black lines.
Streams, blue (or black); double line when over 15ft. wide.
Roads, outlined in black; metalled roads may be coloured brown.
Woods, to be shown by symbols for trees or coloured green.
Villages and *houses*, black.

Scale $\frac{1}{31680}$ or 2 Inches to 1 Mile

Scale of Horiz!. Equiv!ˢ V.I. 25 ft.

VARIATION 16° W.
MAGNETIC

Lithᵈ at the Intell. Div, War Office, Oct. 1902.

Plate VII. To face p. 27.

PHOTOGRAPH OF MODEL OF COUNTRY.

CHAPTER VI.

FIELD SKETCHING.

The Representation of Hill Features.

1. If a model of any part of the earth's surface showing on some reduced scale the hills, valleys, spurs, knolls, cols and ridges, bluffs and cliffs, were to be photographed from above, we should clearly obtain a pictorial representation of the "relief" or hill features of that portion of the earth.

If the light were thrown on the model from one corner—say the top left-hand corner—all the lower and right-hand slopes will be in shade. Such a system has actually been occasionally used in map making, though rarely. In place of photographing a model, we may shade the map (of which the detail has been previously drawn) from our knowledge of the ground, on the assumption that the light comes from the top left-hand, or north-west corner. Such a system is *pictorial shading*, which may be applied with brush and colour, or stump and pencil, or may be used with hachures.

2. Or the sun may be assumed to be vertical, in which case the horizontal surfaces will reflect most light, and the steepest slopes will be the darkest. By an exaggeration of this principle *scales of shade* have in times past been adopted in which the darkness of the shading was a measure of the steepness of the slope.

Such **hill shading** may be carried out with **brush** or **stump**, or by **horizontal** or **vertical hachures**. The hachures are merely broken horizontal or vertical lines, dark and close together where the slope is steep, fine and wide apart where the slope is gentle. The vertical hachures are short disconnected lines drawn directly upward down the slope; the horizontal hachures are short disconnected horizontal lines, that is, lines at right angles to the slope.

The system of **vertical hachures** is used on the following maps:—

> The hill-shaded edition of one-inch Ordnance maps of the United Kingdom. See Plate VIII. Similar maps can be obtained of all important military districts.

The quarter-inch atlas sheets of the Survey of India. The quarter-inch transfrontier and other small-scale maps of the Survey of India, and many small-scale maps published by European states.

It is thus necessary that officers should understand the hachure method of hill shading.

In the maps mentioned no regular scale of shade has been employed, the hachuring gives a good general idea of the larger features, but the slope of any particular part of the ground cannot be ascertained from the map unless (as is the case with the one-inch Ordnance map of the United Kingdom) the hachures are supplemented by contours.

3. There is also the **layer system** of showing hills (*vide* Bartholomew's half-inch maps of the United Kingdom).

4. But for military purposes, by far the most useful system of representing hill features is by **contours, approximate contours, or form-lines.**

A **contour** is the representation on a map of an imaginary line on the ground, joining all points which are the same height above mean sea-level. The high-tide line round the coast is locally, for instance, a natural contour.

In figure 5, p. 29, is shown a perspective view of a simple hill on which the contours are supposed to have been visibly marked. Each of these contours is a line which connects all points on the hill-side which are at the same height above mean sea-level. Thus the contour marked 200 is everywhere throughout its length exactly 200 feet above mean sea-level.

A plan or map of the same hill is shown below. For small areas the plan of each contour may be imagined to be the outline of the intersection of the hill with a horizontal plane. It is clear that, knowing the vertical interval between any two contours and their positions on plan, we also know the mean slopes between them.

The advantages of contours are:—

1. They show the hill forms with exactness, and not vaguely as is the case with shading and hachuring.
2. They require but little artistic talent to draw.
3. They obscure the detail less than shading or hachuring.
4. They can be reproduced by lithography in the field.

ORDNANCE SURVEY

Scale.- 1 Inch to 1 Mile *(Coloured)*

Plate VIII
to face p 28.

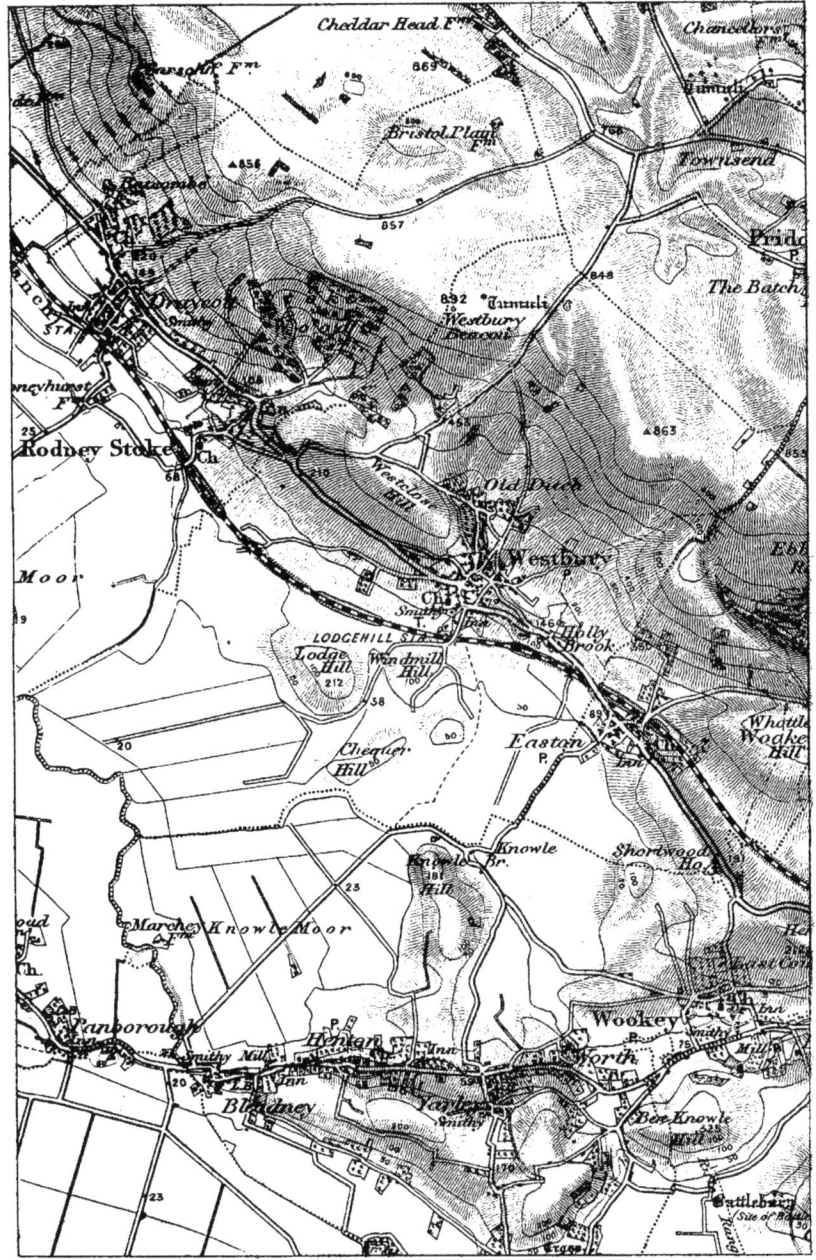

SHOWING THE COMBINATION OF CONTOURS AND VERTICAL HACHURES

The accuracy of position of a contour on a map should be as great as that of the detail, and this is the case in deliberately executed maps such as those of the national surveys. For such accurate maps the contours are determined on the ground by a system of instrumental levelling. This is, however, undesirable on a military sketch.

Fig. 5.

View of a hill with contours supposed to be visibly marked.

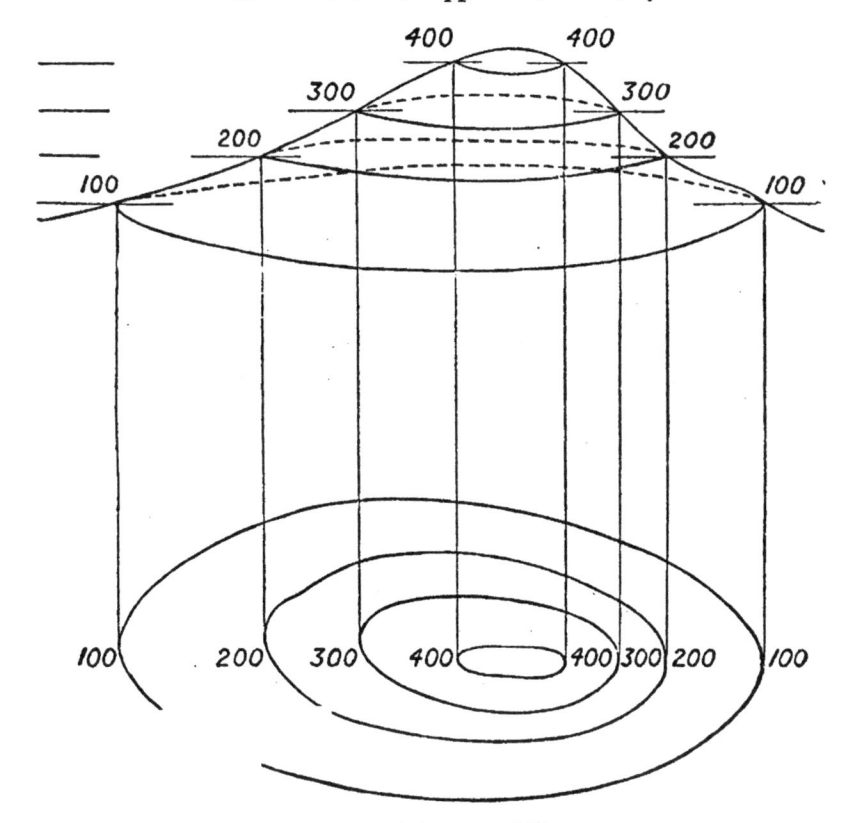

Plan of the same hill.

On military sketches, therefore, hill-features are represented either by **approximate contours**—which are contours determined by approximate methods—or by **form-lines**, which are contours sketched by eye, serving to show the shape of the hill-features rather than their altitudes, but giving at the same time a general idea of the varying slopes of the ground.

Plates XII and XIX are military maps in which the hill-features are shown by approximate contours.

Plate X is an example of the use of form-lines.

There is no very rigid distinction between these two systems; one merges into the other. On very small scales, such as one-eighth inch and one-quarter inch to a mile, the features must necessarily be generalised, and form-lines are used. On half-inch, one-inch and two-inch scales there should be less eye-sketching, and the ground forms should be represented by approximate contours.

The method of drawing approximate contours in the field is described on p. 47, which applies to all the systems in use in making a military sketch; but a few notes are here added on points not dilated upon there.

Vertical Intervals.—The vertical interval between two successive contours is usually expressed in feet, and written V.I. Thus a note on a map "contours at 50 feet V.I," means that any two successive contours are separated by a vertical interval of 50 feet.

Horizontal Equivalent.—This is the horizontal distance, or distance on the map between two adjacent contours; supposing the vertical interval to be fixed, the horizotal equivalent clearly depends on the mean slope of the ground.

The horizontal equivalent is usually expressed in yards, and is written H.E.

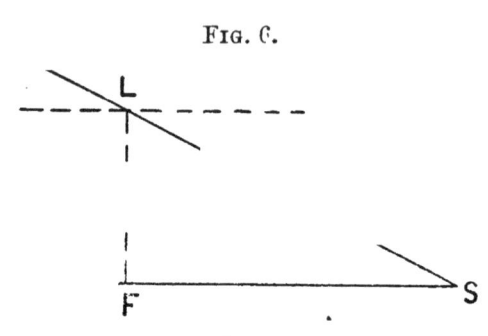

FIG. C.

In the attached figure if S L is a slope, L F is the vertical interval, and F S is the horizontal equivalent for that particular slope and interval.

For a slope of 1° a vertical interval of 1 foot gives a horziontal equivalent of 57·3 feet, or 19·1 yards.

For a slope of 2° and the same rise of 1 foot, the horizontal equivalent is approximately half this, or 9·6 yards; for a slope of 3° it is one-third, and so on. Thus for a vertical interval of 1 foot we have the following list of horizontal equivalents:—

	Yards.
1°	19·1
2	9·6
3	6·4
4	4·8
5	3·8
6	3·2 and so on.

For a vertical interval of 25 feet it is merely necessary to multiply these horizontal equivalents by 25, for V.I. of 50′ by 50, for a V.I. of 100′ by 100.

The following is a *table* of *horizontal equivalents* in yards for a *vertical interval of* 100 *feet*.

Degree of slope.	H.E. in yards.	Degree of slope.	H.E. in yards.
1°	1909	11	171
2	954	12	157
3	636	13	144
4	477	14	134
5	381	15	124
6	317	16	116
7	271	17	109
8	237	18	102
9	210	19	97
10	189	20	91

25°	71
30	58

This table is not meant to be learnt by heart.

A very useful approximate rule is the following:—

$$\text{H.E.} = 19 \cdot 1 \times \frac{\text{V.I.}}{\text{D}}.$$

where H.E. is the horizontal equivalent in yards.
V.I. ,, vertical interval in feet.
D ,, degree of slope (angle of elevation or depression).
Example.—What is the horizontal equivalent for an 8° slope, the vertical interval being 25 feet?

$$\text{Here H.E.} = 19\cdot1 \times \frac{25}{8} = 60.$$

It remains to deal with the practical application of these principles.

In the past, when large scale maps were in vogue, and the 6-inch sketch flourished, a system of normal intervals was adopted; in this system the vertical interval varied inversely as the scale; and on every scale the same slope was represented by contours at the same actual distance apart on paper.

This principle is undeniably sound; but the particular number adopted was not so satisfactory, and the intervals were not in accordance with those actually or tacitly adopted by the great surveys of the world.

The rule which is to be followed in future is this:—

$$\text{Vertical interval in feet} = \frac{50}{\text{number of inches to a mile (of the scale)}}$$

Thus the vertical interval for a map on a scale of 2 inches to 1 mile is $\frac{50}{2}$, or 25 feet.

This rule may be departed from in those cases in which large-scale sketches are required. In such cases officers may at their discretion use larger intervals. By large scale is here meant a scale of more than 2 inches to 1 mile.

This rule, like all rules, should be applied with common sense. For instance, suppose it is required to make a local sketch of a village on the 4-inch scale, then the V.I., by the rule, would be $12\frac{1}{2}$ feet; as this is inconvenient, use 10 or 15 or 20, preferably 10 or 20, for the reason which follows.

The 1 inch Ordnance maps of the United Kingdom show contours at intervals of 100 feet; in any operations at home these maps, or enlargements of them, will be freely used; it is therefore necessary that sketches made for use, in conjunction with the Ordnance maps, should show the 100, 200, 300 . . . contours. The same applies to the use of the new ½-inch Ordnance map. It will be seen that the above rule gives the required contours on 2-inch, 1-inch, and ½-inch scales.

This rule has been adopted for the topographical maps of South Africa; it is that on which modern Ordnance survey topographical maps are made, such as the maps of Jersey and Mauritius; and it is also that tacit rule on which the form-lines of the Indian frontier maps are sketched in. The effect of this system is that the contours are much closer together than on the old system, and on steep slopes it has somewhat the effect of a shade; the hill features in consequence show up well and require no addition in the shape of artificial shading. The great objection to any system of stump or brush shading is that a map so shaded cannot be reproduced in the field, whereas any system of contouring can be so reproduced: many thousands of topographical maps contoured on this system were printed on lithographic presses at Orange River, Bloemfontein and Pretoria during the South African war of 1899–1902. Such contours should be drawn in brown (not red, which dazzles the eye and takes away from the shade effect): the best brown to use is burnt sienna. If brown is not available, red may be used. (Plates VI, X, XIX, show maps so contoured.)

In rapid sketching in the field it often happens, however, that there is no time to colour the pencil sketch. In such a case the contours should be shown as chain-dotted pencil lines (see Plate XII). When chalk pencils are available, the contours may be drawn in as brown chalk chain-dotted lines.

Recapitulation.—Hill features may be represented on a map by one of the following methods:—

1. Pictorial shading with brush or stump, or hachures.
2. The layer system.
3. Horizontal hachures.
4. Vertical hachures.
5. Contours.
6. Approximate contours.
7. Form-lines.
8. Or by combinations of these, such as vertical hachures and contours, as in the 1-inch hill-shaded Ordnance maps.

The system adopted for military maps is that of approximate contours or form-lines.

For ordinary military scales of 2 inches to a mile and under, the vertical interval of the contours, in feet, is to be

$$\frac{50}{\text{number of inches to a mile}}$$

The contours are to be drawn in brown. When paint is not available, the contours are to be shown as chain-dotted pencil or chalk lines.

Officers should use their discretion as to the vertical interval to be used on large scales.

Scale of horizontal equivalents, to be used with scales of two inches to one mile, and all smaller scales.

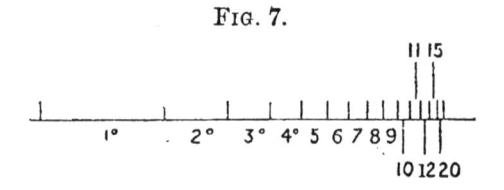

Fig. 7.

CHAPTER VII.

FIELD SKETCHING.

Instruments used in Field Sketching.

The following are the instruments most likely to be available for field sketching.

      Plane-table; for description and use, see p. 38
      Prismatic compass   ,,   ,,   ,,   50
      Abney's level   ,,   ,,   ,,   36
      Cavalry sketching board   ,,   ,,   70
      Field book   ,,   63
      Aneroid barometer   37
      Range-finder   ,,   ,,   73

It is not intended to imply that all these are required for any one sketch.

A field-glass is always useful in field sketching.

As to which of these instruments should be used on any particular occasion, officers on service will generally have to use those which they can get and not those they would wish. The following will, however, give an idea of the relative advantages of plane-table and compass.

The relative advantages of Plane-Table and Compass.

With care, the positions of important points on a plane table may be determined within the thickness of a fine pencil line. A prismatic compass bearing is burdened with two sources of error, one of about $\frac{1}{2}$ degree in the observation, and another of, say, $\frac{1}{4}$ degree in the plotting. On the score of accuracy, in favourable country, the plane-table is much to be preferred to a compass.

Again, the plane-table, properly used, does not depend at all on the compass and the results are not affected by local magnetic attraction. There are many parts of the world where a prismatic compass sketch is rendered valueless by local attraction.

Again, hill sketching is far more difficult with a compass than with a plane-table. For an area sketch it is far quicker to use a plane-table.

Against this may may be put the facts that:—

The plane-table ceases to be an effective instrument in very flat country, in high grass, or in thick forest, as it is then impossible to work by the methods of intersection and resection.

The plane-table cannot be used on a river or on horseback.

It follows that for area sketches, and reconnaissances of defiles, passes, outpost and defensive positions in open or fairly open country a plane-table should be used in preference to a compass. The compass should be used in forest country, in high grass, for a river or road reconnaissance.

The **Abney** level. This consists of a rectangular metal tube ab (Fig. 8) about 4·5 inches long. At one end is an eye-piece E, at

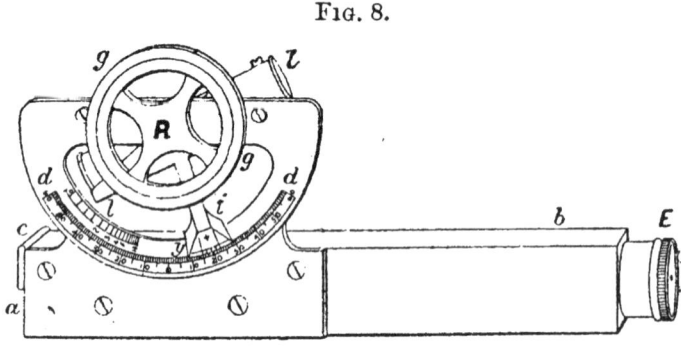

FIG. 8.

the other is inserted a hollow tongue of metal c, which carries a silvered plate inclined to the optical axis at an angle of 45°. In some patterns the metal piece is replaced by a horizontal crosshair. A slot cut through the upper edge of the tube ab allows the spirit-bubble when in the centre of its run to be seen from E, by reflection, at an angle of 90°.

The spirit level l is attached to the index arm i, which moves over the graduated arc dd. It is capable of movement in a vertical plane round the centre R, and is placed immediately over the right-hand edge of the mirror in the tube, so that when the index is at zero any object which coincides with the centre of the bubble, seen by reflection from E, will be at the same level as the observer's eye.

The arc is graduated to degrees, the vernier on the index arm enabling angles to be read to 10 minutes. Those who cannot read a vernier scale can judge the position of the zero arrow to the nearest half degree. This is usually accurate enough for field sketching.

For convenience of reference, the slopes between $\frac{1}{1}$ and $\frac{1}{10}$ are engraved on the arc opposite to the number of degrees which they represent. In reading this scale the edge y of the index arm, and not the zero of the vernier, must be brought into coincidence with the several divisions.

To observe the angle of elevation or depression to any object, the instrument is held in the right hand and the line of sight directed through the right-hand portion of the tube in the required direction. The centre of the bubble is then made to coincide by reflection with the object by moving the milled screw g, the required angle being shown by the index arrow. A horizontal wire is fixed in the centre of the tube to facilitate this adjustment.

THE ANEROID BAROMETER.

Aneroid Barometers.—For military work a $2\frac{1}{2}$-inch barometer reading to 6,000 feet is a useful pattern.

In settled weather, and in continental climates, the aneroid barometer is often useful in sketching, especially on small scales where the contour intervals are large. For instance, in sketching on a scale of 1 inch to 1 mile, where the intervals will be 50 feet, or on the $\frac{1}{2}$-inch scale with intervals of 100 feet, the aneroid will be found most useful. The aneroid should not be used in unsettled weather when the atmospheric pressure is varying.

An aneroid used for field sketching should be provided with a movable scale of heights in feet. If starting from a point of which the height is known, the scale should be set to that height before beginning work, or else a plausible height should be assumed; the reading should also be taken on returning. If there is a difference, the intermediate altitudes should be corrected in proportion to the time which has elapsed.

An aneroid should always be read in the same position, either always held vertically on a level with the eye, or always horizontally, the latter is better. Severe tapping is not good for an aneroid barometer.

A very rough rule is that a difference of height of $\frac{1}{10}$ inch indicates a difference of level of 100 feet.

A closer approximation is :—

If H be the difference of height in feet of two stations.

S the sum of the aneroid readings in inches, tenths and hundredths.

D the difference   ,,   ,,   ,,   ,,

$$H. = \frac{52,000 \, D}{S.}.$$

For accurate methods of computation, see "Text Book of Topography."

In changeable weather barometers are, of course, quite unreliable.

Diurnal variation of the barometer. This variation, which is not very large, depends on the place and time of year. On account of this variation it is desirable, if possible, to arrange for checks on the barometer readings—usually impossible, however, in field sketching.

Range finders. See Chapter XI.

THE PLANE-TABLE.

The **plane-table** is merely a portable table. The table or board can be carried separately from the stand or legs; the stand is usually a folding tripod. For greater compactness, the stand is sometimes made of a collapsible pattern. The adjuncts to the plane-table are the sight-rule and the trough-compass; the plane-table should also be provided with a waterproof cover.

There are several patterns of plane-table, suited to various conditions. Generally, plane-tables used on a rigorous deliberate survey will be heavier and larger than those used for field sketching. The large Indian plane-table, for instance, is a board 30 inches by 24 inches, which is larger than the military topographer could conveniently carry. Where the plane-tabler uses a horse to get about on, as is customary in South Africa (though of course he has to get off to use the plane-table), it is very desirable that the plane-table should be of a light, collapsible pattern. There is a Service pattern of plane-table which is 18 × 24 inches, and which has hitherto

Plate IX. To face p. 38.

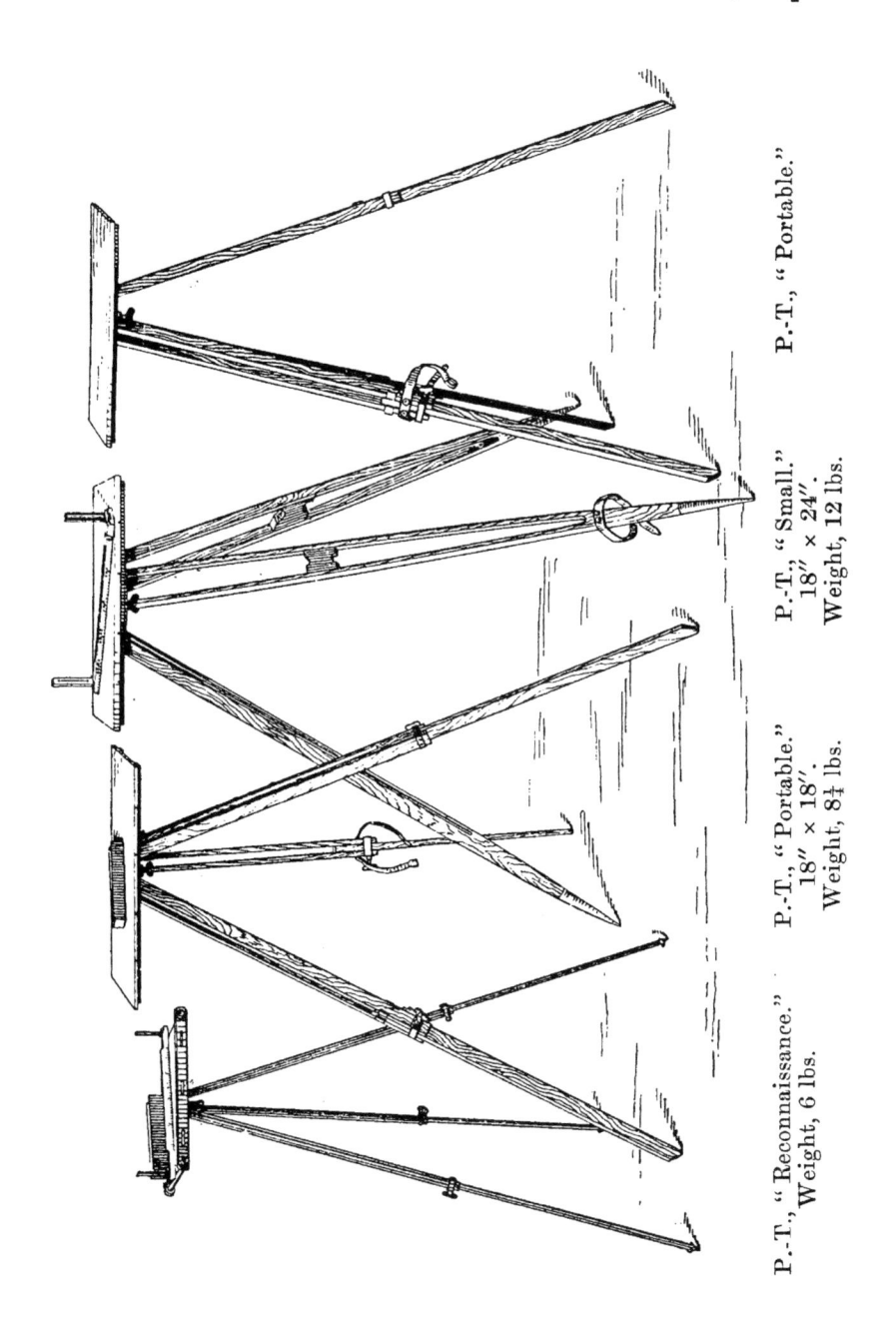

P.-T., "Reconnaissance."
Weight, 6 lbs.

P.-T., "Portable."
18" × 18".
Weight, 8¼ lbs.

P.-T., "Small."
18" × 24".
Weight, 12 lbs.

P.-T., "Portable."

been used by the survey sections, R.E. There is also a Service "reconnaissance plane-table," provided, like the cavalry sketching-board, with rollers; this takes paper 11½ inches wide, and has brass, tubular legs. This is a useful instrument for special conditions, and during the operations at Paardeberg an officer made an excellent plane-table sketch with this instrument; but the tubular legs easily get dented and unworkable, and the table is too small for general military use. There is also the service "portable" plane-table, 18 × 18 inches, with wooden "camera" legs, which is perhaps the most generally useful.

Illustrations of these various plane-tables are given in this book (Plate IX). Officers need not consider themselves tied down to the use of any special form of plane-table.

The **sight-rule** should be about as long as the shortest side of the table, and the fittings should be of gunmetal or brass, not iron or steel. A common defect of sight-rules is that the slit in the rear vane is too small. When working in hilly country the centres of the tops of the vanes should be joined by a string, tense when the vanes are upright, to enable rays to be taken on steep slopes. It is convenient, as in the Service pattern, for the sight-rule to be graduated with useful scales, notably 2 inches to 1 mile.

The **trough-compass** should not be combined with the sight-vane. The 6-inch trough-compass in mahogany box with sliding lid, which throws the compass off its pivot, is the best. Trough-compasses should be stored horizontally. They can be adjusted for "dip" by tying string round the elevated end and dropping sealing-wax on to the string. Where a trough-compass is not available, the Service prismatic compass will do instead.

Mounting a Plane-table.—With an ordinary plane-table which has a plain board (and all arrangements of clips should be eschewed), if time presses, the only thing to do is to pin the paper on to the board with drawing-pins, and if the paper is linen-backed so much the better.

But if time allows, "mount" the plane-table as follows :—

Get some fine linen about 8 inches larger in each direction than the board, damp this slightly, and paste it on to the board on the underside only. The paste should be made of the finest flour available; cornflour is the best. Now take a piece of drawing-paper the same size as the linen, damp it on the wrong side, paste the linen all over and smooth the paper down on to the linen;

paste also the overlap of paper on the underside of the table. Put in a few pins temporarily on the underside; these can be removed in about 12 hours, when the table will be ready to work on.

When the work is finished and the field sheet is removed from the board, its edges should be bound with tape.

The reasons for mounting paper on linen are that its renders the paper less liable to be torn, and somewhat less liable to be affected by expansion caused by wet and damp.

SURVEYING AND SKETCHING WITH THE PLANE-TABLE.

The plane-table is by far the most useful instrument employed in making military maps and sketches, and it is very important that the student should master the methods here described. The following are **general maxims** which should always be observed:—

1. See that the leg-screws and clamping-screws are taut before beginning work.

2. Spread the legs widely, especially with collapsible patterns; the stiffness of the table largely depends on the spread of the legs. In soft soil the legs should be well pressed into the ground. The board should be horizontal.

3. Always keep the pencil sharp; there should be no error in essential features greater than the thickness of a pencil line. With acute intersections, which must sometimes be used, fineness of line is absolutely necessary. Points required afterwards for interpolation should be fine dots surrounded by a circle. No good results can be expected with a blunt pencil.

4. When it is required to draw a ray from a point, put the pencil upright on the table with the edge of the base just touching the point, and lay the sight-vane against the pencil. Or instead of a pencil some expert plane-tablers use the finger. *Pins would never be used by a good plane-tabler*.

5. When drawing a ray from a point, set the point of the pencil into the point first, so as to get the correct angle at which the pencil should be held when drawing it along the sight-vane. Without this precaution large errors creep in.

6. Whenever the plane-table has to be set up or oriented by a previously drawn short ray, it is most important that the ray, when it is first drawn, should be extended by marking short

lengths of it at or near the edges of the board. These are called
"repere" marks and give a good line along which to lay the
sight-vane.

7. When drawing a ray to a distant object, mark its approximate position, as estimated, by a small circle on the ray. Write the name of the object along the ray; to assist identification it is sometimes useful to draw a small sketch of the object at the end of the ray.

In an extended survey, plane-table work is based upon a triangulation executed with theodolites; but for a military sketch it is not usually possible to carry out such a triangulation, and very excellent results can be obtained without it. Even in an extended survey where time is of importance an instrumental triangulation is sometimes dispensed with, as in the case of an excellent ¼-inch survey carried out in 1891 between the rivers Salween and Mèkong. It will therefore be assumed in the following remarks that the military topographer is ordered to make a plane-table sketch of a block of country without the aid of refined instruments.

The steps to be taken in making a plane-table sketch are as follows :—

1. **Select a Base-line.**—This should be on level ground. It should occupy a fairly central position, though this is of no great importance.

Both ends of the base should be visible from well-marked natural points in the area to be surveyed, and from each other.

The length of the base would usually be from ½ mile to 1 mile, but the longer the better. A good rule is that the base should be about 2 inches long on the paper, but this is longer than can usually be obtained. If possible mark the ends of the base, or, if this cannot be done, measure between existing objects, or else measure in the same line as two existing objects.

2. **Measure the Base.**—If neither chain nor tape are available, mark a piece of rope of length about 20 feet (or, if possible, longer) into foot lengths by means of a protractor, or from any scale such as those on the sight-vane, and measure with this. If rope cannot be got, then pace the base. Under any circumstances, if time presses, the base must be paced.

3. **Fix Points by Intersection.**—The most suitable position on the board must be carefully considered before commencing work; the amount of country to be surveyed, and the position of

the prominent points required for interpolation, being taken into consideration.

Set up the plane-table at one end of the base, and draw rays to the other end of the base and to prominent objects within the area to be surveyed, or not far outside.

Place the compass in its box on a corner of the table, and move the compass until the needle points steadily to the centre division; then draw a line on the table at each side of the box, and remove the compass. Check the position of the needle subsequently in the course of the work at intersected stations.

Go to the other end of the base, place the sight-vane along the pencil line representing the base, set up the plane-table by moving it until the sight-vane is aligned on the first end of the base, then draw rays to the same prominent objects used before.

Then go to one of the points intersected which makes a well-conditioned triangle with the base, test its position by setting up by one end of the base and lining on the other, and then draw rays to all the other intersected points which are visible, and to any new points which may be likely to be useful. Except in the

FIG. 9.

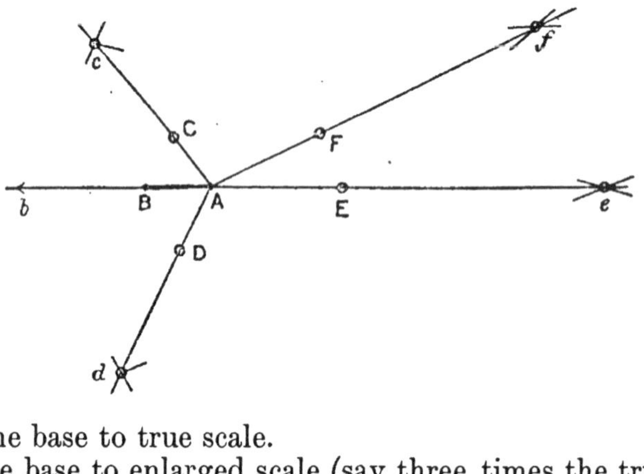

AB is the base to true scale.
Ab is the base to enlarged scale (say three times the true).
c, d, e, and f are positions on enlarged scale.
E and F the positions of e and f on true scale.
So that AE = $\frac{1}{3}$ Ae, AF = $\frac{1}{3}$ Af, &c.
This method will not often be necessary.

immediate neighbourhood of the base, it is desirable that each intersected point should be fixed by three rays, particular care being taken to ensure the greatest possible accuracy in fixing points which may afterwards be used for interpolation. Acute intersections are to be avoided; rays intersecting at right angles, of course, give the most accurate results.

It is by no means necessary to cover the whole of the sketch with intersected points, but it is important that the most prominent points should be so fixed.

If the base is very short, it may be necessary to plot it on a larger scale than that intended for the sketch, reducing it before beginning work when a few fairly distant intersected points have been fixed, and rubbing out the larger scale work. (Fig. 9, p. 42.)

4. **Take Aneroid Readings.**—Whilst visiting the above points take careful readings of the aneroid barometer, having assumed some height as a starting point. The resulting heights should be written against the points. The aneroid heights should not be crowded together, and cannot be relied upon in changeable weather; in this case other instruments must be used to determine heights, *e.g.*, Abney's level.

5. **Fill in the Detail; mainly by Resection.**—Interpolation, or fixing by **resection from three known points**, sounds a very formidable process, but is in reality very simple. It is the principal means by which the bulk of the detail should be fixed. The method is as follows:—

Set up the plane-table at any point from which three previously fixed points can be seen. Orient the table roughly by the trough-compass, and from the three fixed points draw back rays. If these three rays pass through a point, this point is the required position. If they do not pass through a point, the rays will form a small triangle called the "triangle of error." The true position is now to be determined by the following rules:—

I. If the "triangle of error" is inside the triangle formed by the three fixed points, the position is inside the triangle of error; and if outside, outside.

II. In the latter case the position will be such that it is either to the left of all the rays when facing the fixed points, or to the right of them all. Of the six sectors formed by the rays, there are only two in which this condition can be fulfilled.

III. Finally the exact position is determined by the condition that its distances from the rays must be proportional to the lengths of the rays.

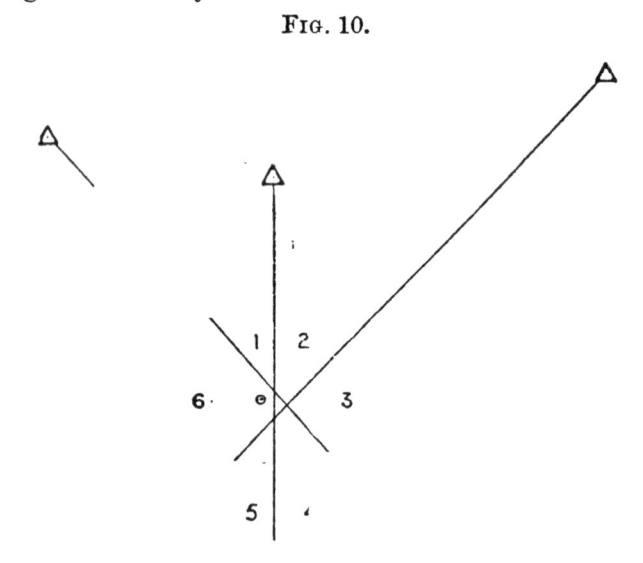

FIG. 10.

In the attached example, for instance :—

By condition I the point must be outside the triangle of error.
,, ,, II ,, ,, ,, in sector 6 or in sector 3.
,, ,, III ,, ,, ,, in sector 6, since the distances from it to the rays must be proportional to the lengths of the rays, and by estimation it will be where shown.

Having thus determined the position, place the sight-rule along the line joining this and the most distant of the points used, set the sight-rule on the point by shifting the plane-table; clamp and test on the two other points. If there is still an error (which should, however, be much smaller), go through the process again.

The best position is inside the triangle formed by the three fixed points, of which two are near and one is distant. Accuracy of *position* is ensured by two points being near, accuracy of *setting* is ensured by aligning on the distant point. And in general fix from near points, set by a distant point.

It is very necessary to be able to make fixings by this method both rapidly and accurately. It is the essential foundation of all good topographical plane-tabling. The plane-tabler in open country need not pace or measure a yard after he has fixed his base. In addition to the speed which results from a free use of interpolation there is also a gain in accuracy, since there is no piling up of errors, and each fixing is dependent only on the main intersected or other well-fixed points.

There is one case in which the method fails, viz., that in which the observer's position and the three points lie on or near the circumference of a circle. The plane-tabler should be on the lookout for this, and not interpolate from points so situated.

It will have been noticed that the trough compass is only used to orient the plane-table roughly, the final fixing does not at all depend on the compass. It is important that this should be remembered.

5. **Resection from Two Points.**—It will, however, sometimes happen that only two fixed points or stations are visible; in this case the plane table must be oriented by the compass, and rays must be drawn back from the two fixed stations intersecting in a point which must be taken as the required position. It is clear that such a fixing depends entirely on the compass, and may, therefore, be very inaccurate. In parts of S. Africa such a fixing would, on account of local magnetic attraction, be entirely worthless, and it is obvious that as a general rule compass fixings should be avoided.

This method should, if possible, only be used for putting in detail, and not for throwing out points on which future work will depend.

Whenever a fixing has been made the local detail should be drawn in; a combination of measurement and estimation should be used; the amount of estimation permissible will depend on the scale. Thus, on scales of 2 inches to 1 mile and smaller scales it is not necessary to pace the width of roads or sizes of houses: these are put in conventionally. On a scale of 2 inches to a mile a distance of 8 yards is less than $\frac{1}{100}$ inch, so that any distance may be estimated which is not likely to be 8 yards in error, and so on. When time presses, minute accuracy must be sacrificed.

6. **Determine the Heights of the Plane-table Fixings.**— A certain number of fairly reliable heights having been determined

by barometer, at each plane-table fixing take vertical angles to one or more of these with an Abney's level or clinometer. It is generally desirable that a height should not be determined from a point which is more than a mile distant, on account of the comparative roughness of the instruments used. But no fixed rules can be laid down.

Example.—At a plane-table fixing the angle of depression to a previously determined point, as measured by an Abney's level, was 1° 15′; the distance measured on the plane-table was found to be 2,920 feet; and the height of the distant fixed point was 310 feet.

Now there is a very convenient rule which all military topographers should remember, *i.e.*, that a slope of 1° is equivalent to a rise or fall of 1 in 57·3 (or, roughly, 1 in 60) and with other angles in proportion; thus 2° in a slope of 1 in $\frac{57\cdot3}{2} = 28\cdot7$, 3°, 1 in $\frac{57\cdot3}{3} = 19\cdot1$, and so on. Therefore an angle of 1° 15′ is equivalent to a slope of $1\frac{1}{4}$ in 57·3 = $\frac{1\frac{1}{4}}{57\cdot3}$ or $\frac{1}{45\cdot8}$. Hence the difference of level of the two points in question is $\frac{2920}{45\cdot8}$ feet or 64 feet. Hence the height of the plane-table fixing in this case is 310 + 64 feet or 374 feet.

7. **Traverse Detail which cannot be Fixed by Resection.**—It will sometimes happen even in open country that there are parts of the work, such as valleys and woods, in which plane-table fixings cannot be made; in this case it is necessary to have recourse to plane-table traversing. Plane-table traversing has to be much used in flat country unprovided with prominent landmarks, and was so used in the mapping carried out during the operations in China in 1901.

The method is as follows: Set up the plane-table at some known point and orient it, draw a ray in the direction of some visible object (and the farther off this is the better), pace to this second point, set up the plane-table there, place the sight-rule along the ray previously drawn, and turn the table until the sight-rule is on the first point; clamp; measure the paced distance along the ray, and mark the second point on the plane-table; draw a ray to some third point, pace to it, and go on as before. Continue the traverse until it is possible to make an accurate, fixing by interpolation, and then adjust the traverse between this fixing and the starting-point. Errors are inevitably generated in this sort of traversing, which should therefore be penciled in quite lightly, as it will have to be rubbed out when being adjusted.

The adjustment is to be carried out as follows :—

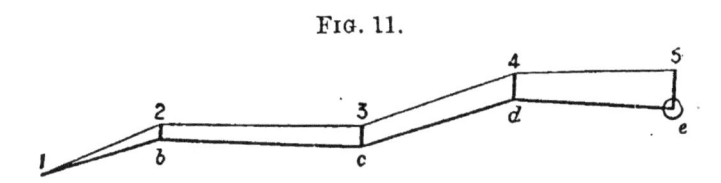

Fig. 11.

If 1, 2, 3, 4, 5 are the stations of a traverse, a fixing was made at 5, and the true positiod of this fixing was e; draw lines through 2, 3, 4 parallel to $5e$, and make $2b$, $3c$, $4d$ bear the same proportion to 12, 13, 14 that $5e$ does to 15. This, except in the case of large errors, can usually be done by eye. Then the traverse as corrected will be 1, b, c, d, e.

Or instead of orienting by back rays as above described, the plane-table may be oriented by compass alone. But it is not advisable to use this method where it can be avoided, as a little consideration will show; for a compass may easily be $\frac{1}{2}°$ out, which is equivalent to a lateral error of $\frac{1}{120}$; now, in aligning by a back ray it is very unlikely that the sight-rule in a ray 120 yards long would be 1 yard off the object. However, where the rays are very short, or the traverse is long, or objects to line back on cannot be found, or where time is short, it may be necessary to traverse with a compass.

It is sometimes convenient to plot the traverse at the side of the board and subsequently to transfer it to its proper position when corrected.

8. **Contour the Sketch.**—We will now suppose that by the various methods of measuring a base, intersection, resection, and traversing, the detail has been drawn on the plane-table, which is also well covered with heights determined during the process of intersection and interpolation. It is now necessary to contour the sketch.

Set up the plane-table at any point of which the height has been previously determined, and orient the table by aligning on some distant fixed point. Now if the ground slopes fairly uniformly, take a clinometer reading of the slope and determine where the first marked contour comes. Thus, suppose the height of the point where the plane-table stands is 563, and the sketch is to be contoured at 50 feet intervals, the first-marked contour down will be 550. If the angle of depression, as determined by clinometer, is 9°, this is equivalent to a slope of $\frac{9}{57\cdot 3}$ or $\frac{1}{6\cdot 4}$; the distance from the plane-table of the 550 contour will therefore be 13 × 6·4 feet, or 28 yards. Draw a ray in this direction and measure 28 yards, and make a dot or mark and write 550. If the slope continues uniform the contours, being at 50 feet V.I., will be at 50 × 6·4 feet H.E. or 107 yards.* From the 550 downwards along the ray measure a series of marks at intervals of 107 yards; these marks will show the positions of the 500, 450, 400 contours, &c. To determine where they stop, take the depression to some point in the bottom of the valley, of which the distance can be measured on the plane-table, and work out its height in the usual way. Do the same all round the point, and join the marks free-hand. But in general slopes are not uniform. The steepest slope of a hill-side is usually neither at the top nor bottom. In this case, if the bottom can be seen and time presses, take the slope and determine the height of the bottom, from that see how many contours there will be on the hill-side and space them by eye, so that they come closer together where the hill is steepest and wider apart where the slope is gentler.

If time allows, however, it is better to take the depression as far as the point where the slope clearly changes, then take another depression, marking the contours as above described. It is to be remembered that in a military map or sketch the slope is of more importance than the actual height of a hill: indeed it is quite possible to contour a map very fairly without being given a single height to start with; but beginners should not try this.

The contours spoken of are more correctly called form-lines or approximate contours, and should be put in when looking at the ground, so as to get the shape of the hill features between the determined slopes. It is important that the watercourses should

* Uniform slopes can of course be measured off from a scale of horizontal equivalents.

PLATE X.

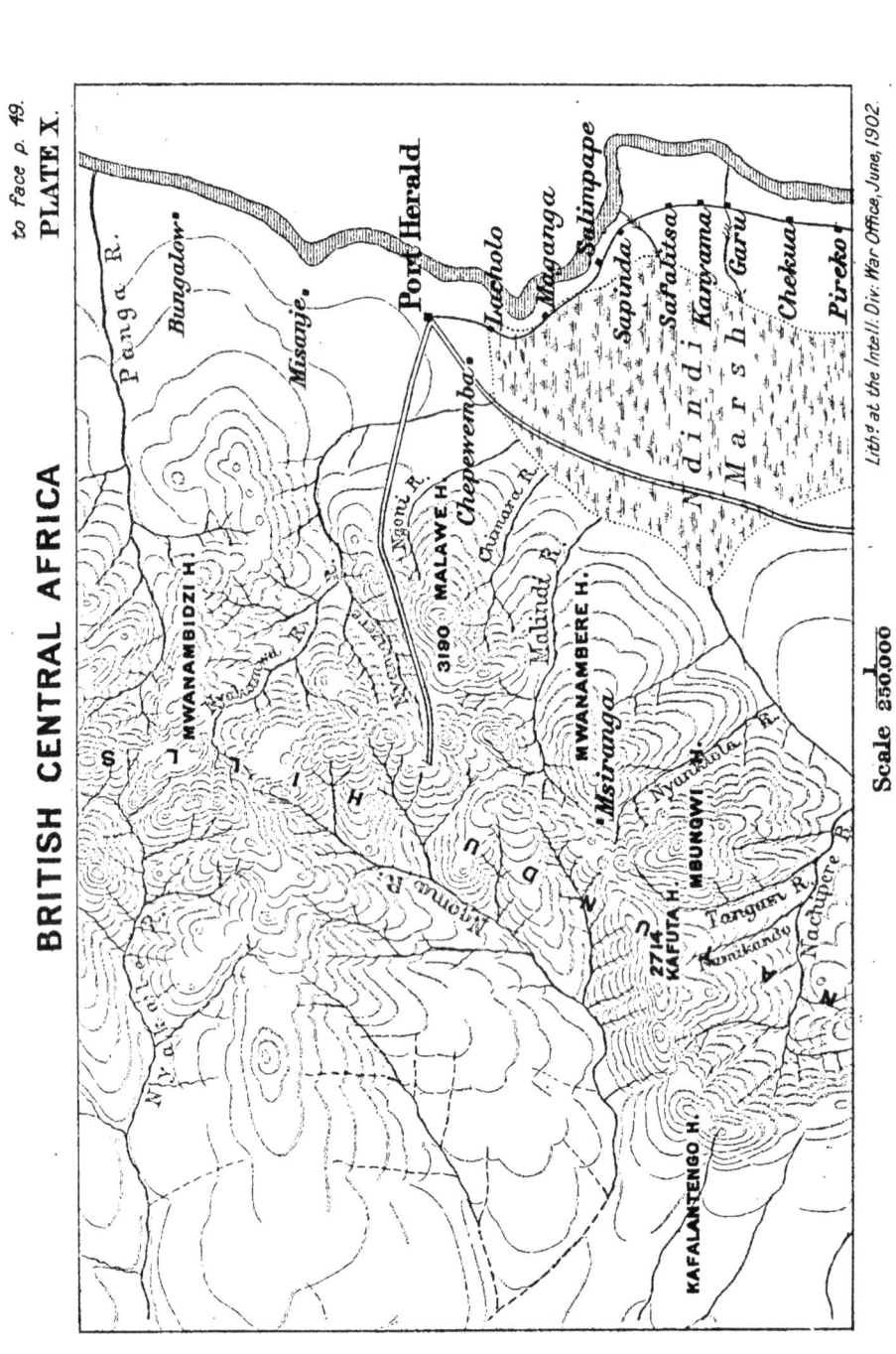

BRITISH CENTRAL AFRICA

Scale $\frac{1}{250,000}$

Specimen of small scale mapping with Form-lines.

Lithd. at the Intell: Div: War Office, June, 1902.

To face p. 4

PLATE XI.

TEXT BOOK OF MILITARY TOPOGRAPHY PART I.

Example of hill drawing, showing the importance of marking the watercourses
Scale, 1 Inch to 1 Mile.

To face p. 49.
Pl. XII.

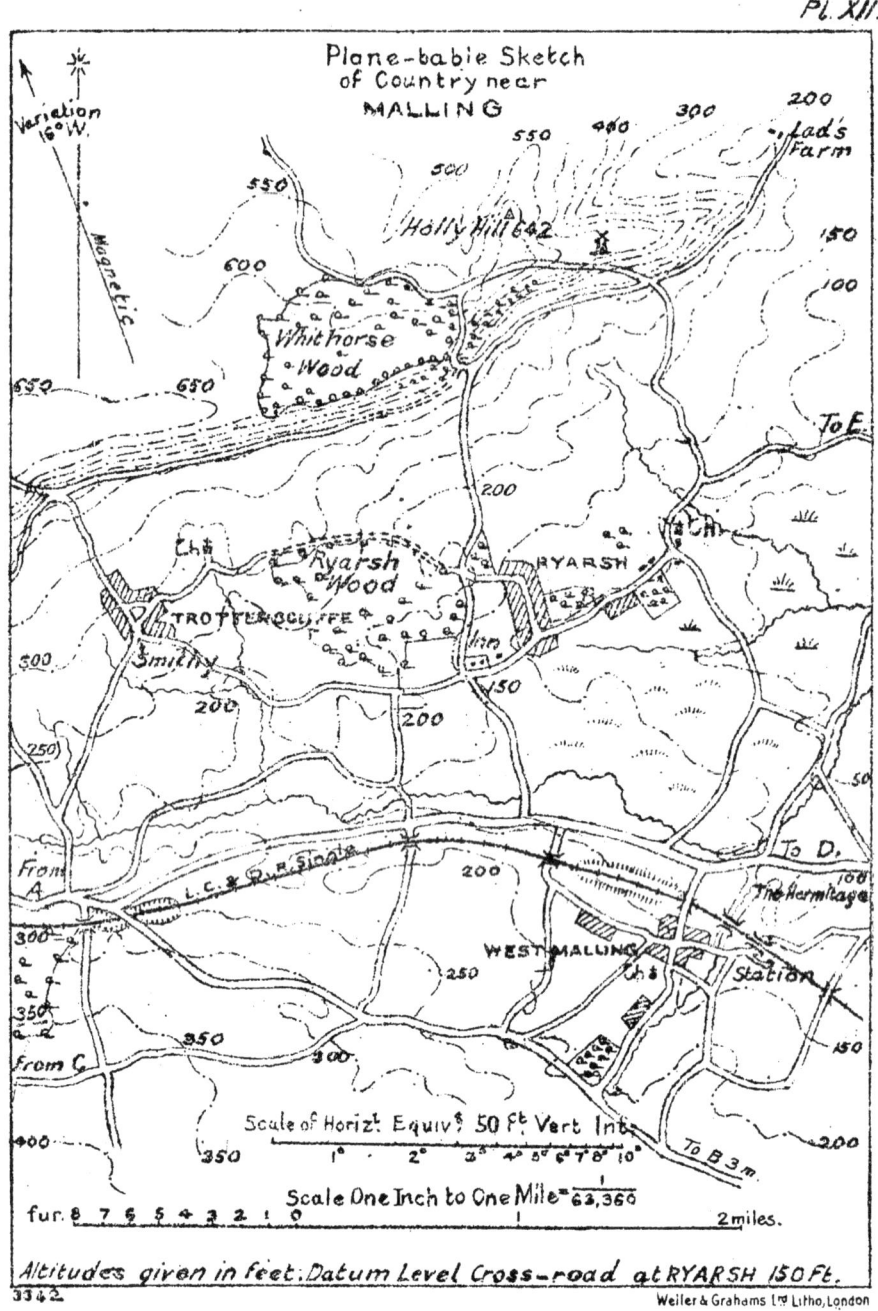

Plane-table Sketch of Country near MALLING

Scale of Horizl. Equivs. 50 Ft Vert Int.

Scale One Inch to One Mile = 1/63,360

Altitudes given in feet. Datum Level Cross-road at RYARSH 150 Ft.

be properly fixed; they are of great assistance in drawing hill features, as the attached example of a 1-inch map of part of South Wales will show. (Plate XI.)

The smaller the scale, the more free-hand may be the sketching, until with very small scales, such as $\frac{1}{4}$ and $\frac{1}{8}$ inch, clinometers can be dispensed with, and the hill features can be sketched by eye, provided that the hill-tops, valleys, watercourses, prominent spurs, and isolated features are properly fixed by intersection and re-section.

On all scales small but important under-features, which do not rise by one contour above their surroundings, should be shown by a form-line.

In the above description of plane-tabling it has been assumed that the detail was finished before the contouring was commenced. The accomplished plane-tabler can, however, do both together: it is merely a matter of experience.

CHAPTER VIII.

FIELD SKETCHING.

The Prismatic Compass.

Next to the plane-table the most useful instrument for military sketching is the **prismatic compass**.

In this the bearings of the card are read through a prism which enables the observer to see the distant object and the reading at the same time. There are many patterns of this. One form is the **Service Prismatic Compass.**

This instrument consists of a magnetic needle balanced on a pivot, and carrying a card divided into degrees, contained in a metal box, fitted with a revolving ring. The metal cover opens on a hinge, and is fitted with a glazed window, on which is traced a fine black hair line, for use as a sighting-vane. Opposite the hinge of the cover is fitted a prism, through which can be read the graduated edge of the card, while at the same time an alignment of the object and the sight-vane on the cover is observed through the slit above it. The prism should be moved up or down in its slot till the figures on the card are properly focussed. A clamping-screw is provided for clamping the needle when not in use, and a "check-spring" for checking its oscillations when observing. A brass ring is attached for convenience in holding it.

The card is "luminous" for night work, the north point is marked with a large diamond-shaped figure. A revolving glass ring is fitted over the compass card, and on the glass is a black direction band, radiating from the centre, at the end of which is a small brass "setting vane," the latter working over an external arc graduated to 360°. On the inside of the cover are two luminous patches, which give a good alignment of the instrument at night when it is held in the hand with the cover wide open. There are two small holes in the brass window edge of the cover, so that, if the glass breaks, a horse-hair can be run between them, and an extemporised sight-vane be utilised.

The *divisions of the card* or ring are read eastward of the meridian, or from left to right, like hands of a watch. North reads 0; 20° E. of north reads 20, and so on.

The prismatic compass gives "bearings" and not "angles." The horizontal angles between any distant objects are obtained by taking the difference of their observed bearings.

Example.—The angle between two points, A and B, is required. Their bearings are observed to be respectively 50° and 110°, 110—50° = 60°, the angle required.

To use the prismatic compass first turn up the sight-vane, or cover, and see that it is upright. Then turn up the prism, and raise or lower it on its slide to obtain distinct vision of the magnified divisions on the card or ring. Now stand facing the object to be observed and, holding the compass by the ring in front of the body, watch the swing of the card. By pressing the "check-spring," which touches its edge, check it when nearly midway in its swing. When quite steady, raise it slowly to the eye, holding it horizontally with both hands, so that the card may not touch the cover. Then observe the object through the slit over the prism, and when the thread is aligned upon it, read the division on the card which is cut by the thread. This will be the bearing of the object. The card must be at liberty whilst the observation is being taken, and the instrument be held quite level.

Practice, and a certain amount of "knack," which is easily acquired, are required to take bearings quickly and accurately. In windy weather it may be necessary to kneel, resting the elbow on the knee to get a correct observation.

The instrument can be used as a plane-table compass by opening the cover wide so that it lies back downwards and flat on the table. An alignment is obtained from the two "nicks," one in the brass ring and the other in the end of the flap on the cover.

The prismatic compass, like other compasses, cannot be relied upon for great accuracy. The presence of iron ore in the ground, or other disturbing influences, may affect it to such an extent as to render it almost useless in some localities. However, in England it will seldom err more than 2°.

The whole of a sketch should be executed with the same compass, as different instruments will vary as much as three or four degrees. The magnetic variation of the instrument being ascertained, the true north can be laid down on the sketch. No observation should be taken with the compass in close proximity to a mass of iron, such as a gate. The effect of the attraction of a railway may be avoided by taking the bearing parallel to it from a point 10 or 20 yards on one side of it.

Sometimes the object whose bearing is required cannot be seen from the observer's position, but can be seen if he advances a few paces, or moves a little to the right or left.

To test the centering of a prismatic compass, stand on a straight line, such as a straight road. Observe the bearing of this line in both directions; these should be the opposite points of the compass. Thus, if one be 271° 15', the other should be 91° 15'. Repeat the experiment on another straight line nearly at right angles to the first.

General Notes on the use of Magnetic Compasses.

The use of the "**points of the compass**" is very inconvenient on land, and readings in degrees are far preferable. Should a compass marked in points be the only one available, it is only necessary to remember that as the 32 points correspond with 360°, each point is worth $11\frac{1}{4}°$. The accompanying diagram (Fig. 12) shows the graduation.

Magnetic Declination, sometimes called the "variation of the compass," is the angle between the true north and the magnetic north. If the needle points east of true north the declination is said to be so many degrees east; and if west, so many degrees west, as 18° W.

The declination is subject to three variations, annual, positional, and diurnal.

Annual Variation.—In London the declination was 11° E. in 1576. It then gradually decreased until in 1660 the magnetic needle pointed to *true* north. The needle continued to move westward until in 1814 it reached its greatest westerly deviation (24° 30' W.). Since 1814 the needle has been moving eastwards, and the declination has been decreasing at the rate of some 7' annually; it is now (1903) about 16° W. in London.

Positional Variation.—The declination also varies according to the position on the earth's surface. (See Plate XIII.)

Besides the variation due to the position of the observer on the earth's surface, there is a **diurnal variation** in the declination of

LINES OF EQUAL MAGNETIC VARIATION 1900
Shewing also the
APPROXIMATE ANNUAL CHANGE IN MINUTES OF ARC.

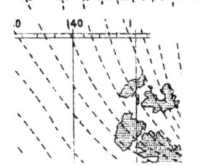

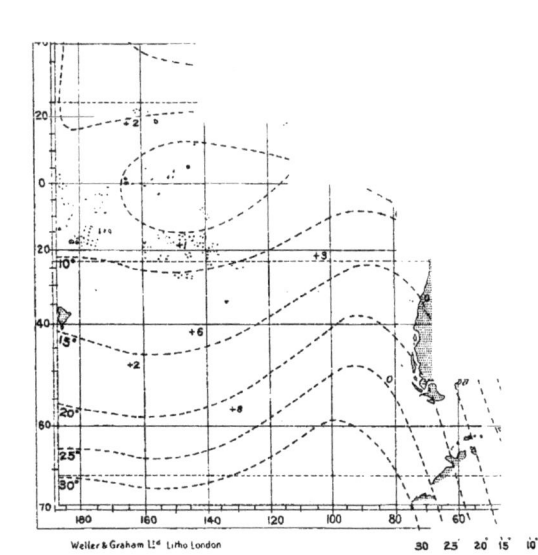

——— Lines of West Variation
- - - - - - " East "

Reproduced, by permission, from "Hints to Travellers" 1900

the needle which differs in summer and winter. In England between sunrise and noon in summer the point of the needle moves westward (*i.e.*, against the sun) for about a quarter of a degree (15 minutes). It then returns gradually until it reaches its original position about 10 p.m., where it remains till morning.

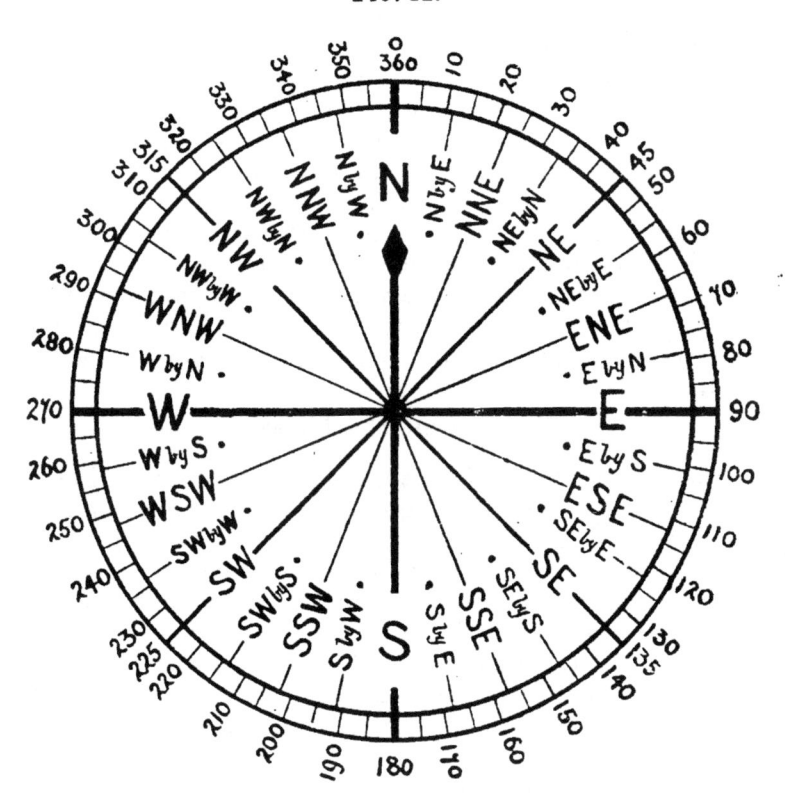

FIG. 12.

The variation is not the same in all compasses with floating cards. This difference is usually due to inaccurate fitting of the magnetic meridian, as shown on the card, to the magnetic needle, or to incorrect centering of the card on the pivot. Hence, when the work of several topographers is to be combined, the compasses should be compared. When making a sketch do not change the compass.

Local Magnetic Attraction.—This is shown by a deflection of the needle from its mean position. It is due to the presence of masses of magnetic iron ore or of iron in the neighbourhood of the needle. For instance, in certain hills in Central Africa a movement of a few feet will cause a movement of the needle of 10° or so. In parts of South Africa there is so much local attraction that it is useless to attempt to work with any form of compass. In a previously unsurveyed country the topographer should always be on the look out for local magnetic attraction.

The determination of the magnetic declination evidently involves the determination of the true north line or meridian. The methods of doing this will be given in the "Text Book of Topography." An obvious approximate way is to take the magnetic bearing of the Pole Star when this is visible. This will not give rise to an error of more than 2° in ordinary latitudes. But in most countries the magnetic declination has been determined and can be found from existing maps, or an approximate value may be used. But it is of no great importance to mark the true north line on a military sketch; it is sufficient to show the magnetic north line and to give the date.

The Dip of the Needle.

A magnetized needle freely suspended by its centre will not hang horizontally except near the equator. The angle which it makes with the horizon is the **dip** or *inclination*. This varies in all parts of the world, being at its maximum at the magnetic poles of the earth, where the needle would point vertically downward. The dip in England is about 70° N., in South Africa about 55° S.

The dip is counteracted by weighting the opposite end of the card or needle, usually by dropping a little sealing wax on a card, or in the case of a needle by tying some thread round it and then dropping the wax.

The Prismatic Compass is to be used in preference to the plane-table, in forest country, in high grass, for a river or road reconnaissance, or where no plane-table is available. Under all other circumstances use a plane-table. The compass should not as a rule be used in countries where there is known to be much local attraction; *e.g.*, South Africa.

Prismatic Compass Sketching.

The drawing is done in the field on a *cavalry sketching board* or on a *sketching case*, which may be simply a piece of stiff cardboard provided with a cover; it is easy to improvise one. The drawing paper is fastened by elastic bands or string, or by pins. The other instruments required are a *protractor*, pencils, knife, and indiarubber.

Dealing first with the protractor. The protractor employed in military sketching is a thin rectangular slab of ivory or boxwood about 6 × 1⅔ inches, with bevelled edges. On it are engraved scales of 6 inches and 8 inches to a mile, and it is also graduated with a scale of degrees radiating from the centre of one edge.

The scale of 6 inches to a mile may be used for sketching on other scales of which it is a measure or multiple, by simply multiplying or dividing the measured distance as required. Thus in sketching at 2 inches to a mile, the measurements on the 6-inch scale would be divided by 3. For instance, 300 yards on the 2-inch scale will equal 100 yards on the 6-inch scale.

For plotting angles the protractor is graduated as below:—

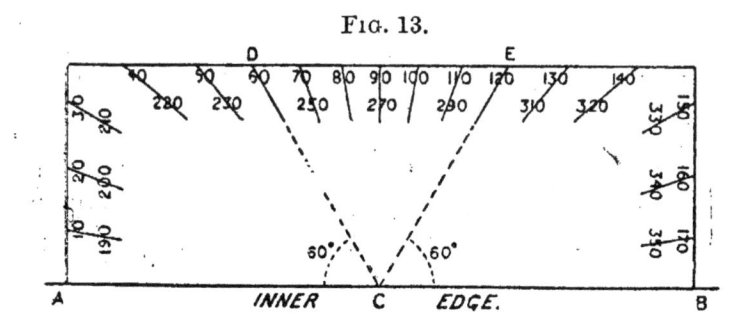

FIG. 13.

Plotting Compass Bearings.

The direction of an object observed with the compass being expressed in degrees from the magnetic north measured round up to 360° with the hands of a watch, is the bearing of the object, and must be laid down on the sketch with a protractor, a more troublesome process than drawing its direction when working with the plane-table.

The sketch must be prepared with parallel lines to represent magnetic meridians, about the width of the protractor apart. The protractor can always be set with its long edge parallel to these lines, by making one of the lines cut each of the shorter edges at similar divisions.

One end of these meridians should be marked N before commencing the sketch, to avoid the possibility of protracting the bearings in the reverse direction.

The protractor is laid on the sketch with its centre at the point whence the bearing is to be drawn, and its long edges set north and south, *i.e.*, parallel to the lines.

If the bearing is under 180°, the graduated edge of the protractor is laid to the right or east; if over 180°, to the left or west, the north margin of the sketch being uppermost.

The bearings for the intersection of stations (Fig. 14) are noted thus:—

At A
{ C, 3°
  D, 40° 15′
  E, 60°
  B, 85° 45′
  F, 110°
  G, 141°
  H, 171° 30′ }

At B
{ F, 181°
  G, 220° 15′
  H, 231°
  A, 265° 45′
  C, 309°
  D, 315°
  E, 350° 30′ }

At G { K, 310°
       L, 66° 15′ }

At D { L, 120°
       K, 242° }

At F — C, 322° 45′

A point is chosen for A in such a part of the paper that there may be room for the whole sketch on it (see Fig. 14).

The protractor is adjusted, with its centre at this point and its graduated edge to the east. The positions of all the bearings under 180° are marked off, and fine lines of indefinite length drawn through them from A, as shown; those over 180° are next protracted.

The measured length of the base being taken along AB, according to the scale of the sketch, the centre of the protractor is set at B, and the observed bearings protracted.

There is no necessity to draw lines throughout from B to intersect the relative lines previously drawn from A. It will be sufficient to mark the points where the intersection will take place.

The letter or name by which each station is designated should be marked on the relative lines as they are protracted, in order to avoid confusion. Sometimes a small sketch may be put at the end of the ray.

FIG. 14.

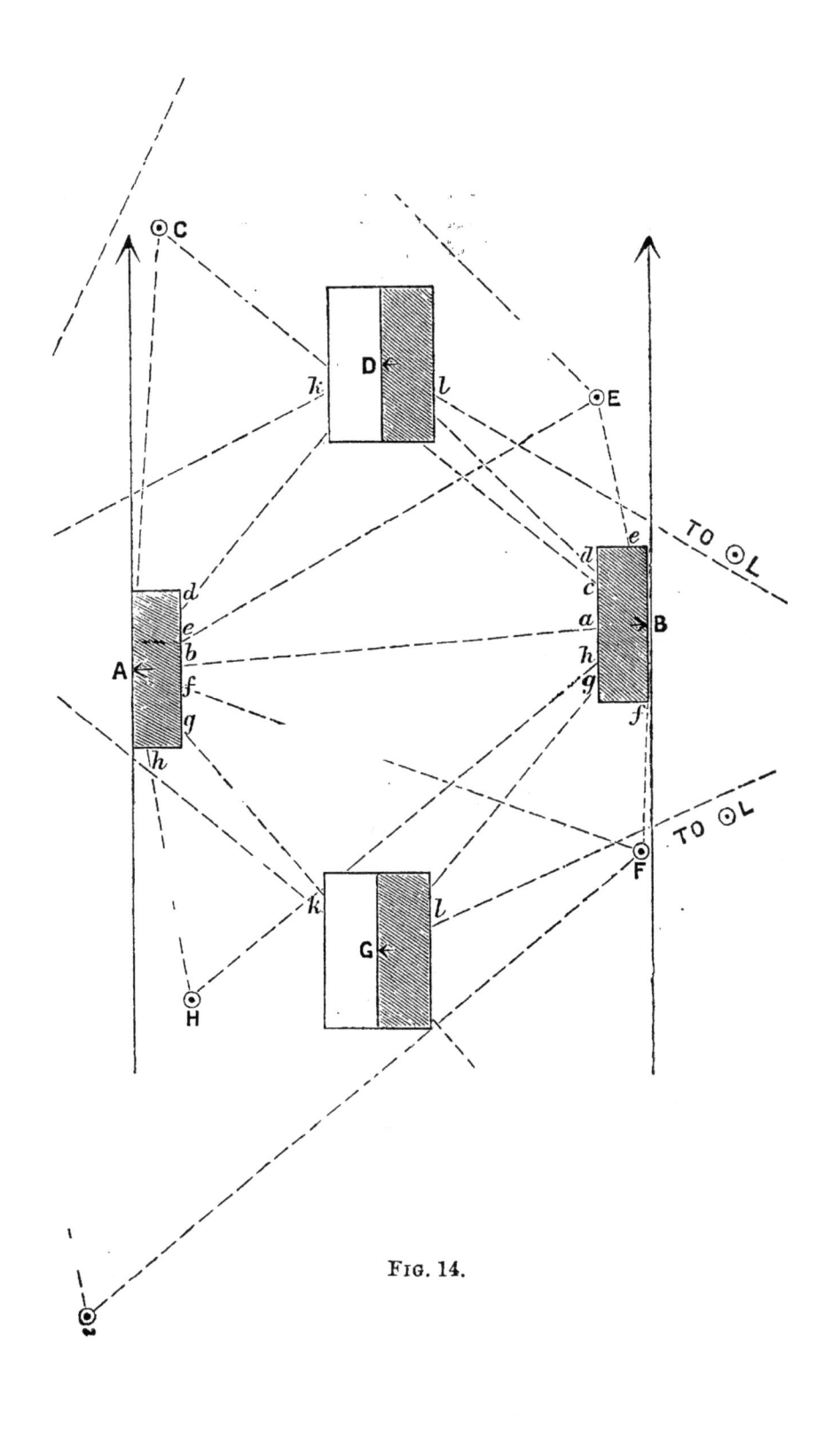

The observation of C at F was taken as a partial test of the accuracy of the former intersection. If this be correct, the line **FC**, when protracted, will pass through C in the sketch, which point was previously determined from A and B.

The point at which the bearing is to be laid down may be so near the edge of the paper that the graduated edge is off it. In this case lay the protractor with the graduated edge inwards, and produce the required bearing outwards *through* the centre.

RESECTION.

Resection with the Prismatic Compass.—The position, x, of an observer (Fig. 15) can be easily found on a sketch by taking the bearings of two or more points already laid down, and plotting the bearings from them in a reverse direction. This process is termed *resection*.

If, for instance, an observer at x finds the bearing of a station, A, to be 45°, then the position of x on the sketch is such that if the bearing of 45° were plotted from it the line would pass through A.

The simplest way of laying down this line is to protract 45° *at A in the reverse direction* (with the graduated edge of the protractor to the left instead of the right), as shown in Fig. 15.

Supposing the bearing of the other station B to be 320°, this would be laid off at B, but reversed, or to the right.

The intersection of these two lines determines the observer's place x. If the intersection be good—*i.e.*, neither too obtuse nor too acute, reliance may be placed on the accuracy of the resection. If the intersection be bad, the position of x should be checked by a bearing from a third point. Indeed in general all points should be fixed by the intersection of three rays.

If the observer at x takes the bearings of three stations, and these bearings, when plotted in the reverse direction, form a triangle instead of meeting, the error should be distributed by assuming the centre of the triangle to be the true position of x, or by using the intersections of the two lines which are believed to be the most correct.

The process of resection is of the greatest use to the sketcher; for if he starts with a few known visible stations correctly laid down, he can resect his position at any point, sketch in the detail near it, and then proceed to another.

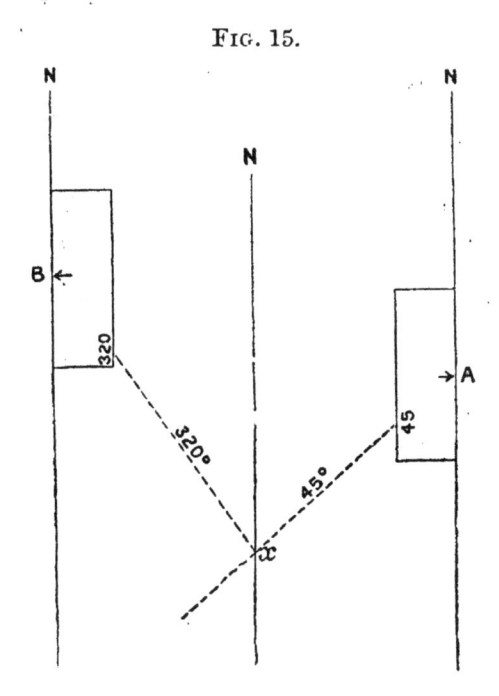

Fig. 15.

It is often useful for finding the position of a cross road, or other convenient starting point, and for checking the accuracy of a traverse.

We will now suppose, by way of example, that it is required to make a prismatic compass sketch of an area of some ten square miles on a scale of 2 inches to a mile.

First rule the paper with north and south lines as described on p. 56.

Next select a base line from the extremities of which prominent objects may be intersected. Working on this sale some care will have to be taken to get a base sufficiently long; *with care* a base half a mile long will do, but a longer one would be better. **The general rules for selecting a base** are as follows:—

(1.) It should be fairly central; but this condition may be dispensed with.
(2.) Its site should be fairly level and free from obstructions.
(3.) Both ends should be visible from the nearest stations, and when practicable should command a good view over the adjacent country.
(4.) When possible the base should not be less than 2 inches long on the paper.

Now as regards the last condition it is clear that it cannot always be fulfilled when working on small scales. This difficulty can be sometimes got over by using a bent base, *i.e.*, a small portion of a traverse (see Chapter IX) with one or more bends in it, such as would be met with in pacing along a road. At each bend the bearings of prominent objects would be taken and plotted.

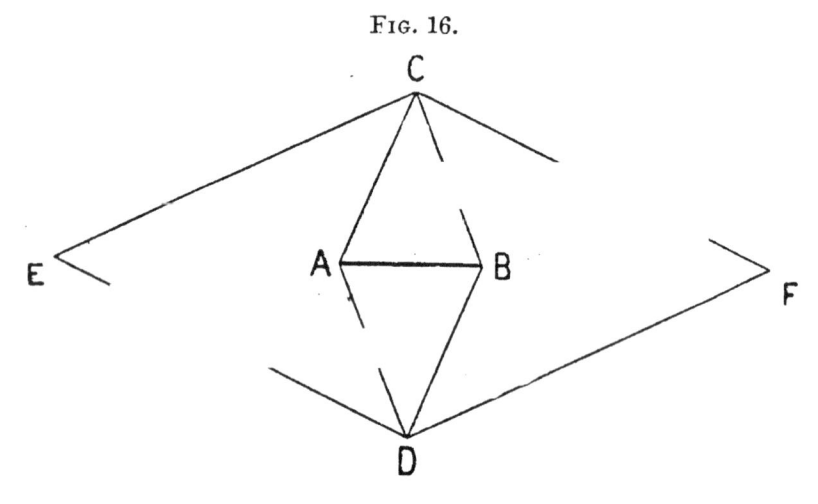

FIG. 16.

But whether the base be bent or straight, the leading principle of fixing points from it and of extending the compass triangulation is the same; and this is that the intersections of the rays should be neither very acute nor very obtuse. The attached diagram shows how a compass triangulation may be extended from a base A B. It is clear that the points E and F could not have been directly fixed by intersections from A and B.

Ordinarily the base would be *paced*. It is best to pace yards.

The length of the pace must be sensibly reduced in pacing down slopes, and increased in pacing up them.

A fruitful source of error in sketching is to lose count of the number of hundred yards which have been paced. Some system of recording these should be adopted, such as closing a finger for each hundred.

If the base is measured along a road, and a *bicycle* provided with a cyclometer is available, it should be used for the measurement; and indeed in a civilized country a bicycle should be used throughout the making of the sketch.

During the process of measuring a base and extending the compass triangulation the local details should be sketched in, partly by measurement and observation, partly by estimation. On the 2-inch scale it is not necessary to pace the widths of roads; it is sufficient to note against them *metalled* or *unmetalled*, and to state the approximate width of the metalling, thus 10′ m., and to draw them in with fine double lines.

It is not necessary on the 2-inch scale to show any hedges, walls, or fences. Nor is it necessary to pace the dimensions of houses; they should be put in their correct positions and be shown as small conventional blocks.

The detail which should be shown on the 2-inch scale is:—

Roads and milestones, rivers, streams, canals, railways, churches, solitary houses and farms, woods, ponds, lakes, swamps, and marshes, wells, smithies, bridges and telegraph lines, and post and telegraph offices. Orchards, where they are rare, should be shown. Villages should be blocked in so as to show their outline and the main roads going through them. The direction of the stream of a river should be shown by an arrow. Where the margin cuts a main road put "from ——" if the place is at the bottom or left of the sketch, and "to ——" if at the top or right.

In ordinary English country it is possible to fix nearly all the main features by intersection and resection. Between these the detail must be filled in by traversing, pacing, and estimation according to its importance.

In steady weather the barometer should be freely used to determine the heights of important points. The slopes, if time allows, should be determined by clinometer. The vertical interval between the contours on the 2-inch scale is 25 feet.

The method of drawing the approximate contours is the same as that described on p. 47.

For conventional signs, see Plates IV, V, VI.

As regards the relative advantages of plane-table and compass, see p. 35.

Good rough area sketching can be done on a cavalry sketching board on foot (see p. 70). The principles of selecting a base, and intersection of stations are the same as just described, but the work, though quicker, is much less accurate. Primarily, however, the cavalry sketching board is intended for linear and not area sketching.

CHAPTER IX.

FIELD SKETCHING.

Traversing and the Field Book.

A traverse is a continuous series of measured straight lines of which the directions have been observed.

The straight lines, called *traverse lines*, are measured with a chain or tape, or by pacing, or by cyclometer, pedometer, perambulator, or by time, or by estimation.

The position of the detail is determined either by the measurement of short lines, called *offsets*, which are perpendicular to the traverse lines, or by taking bearings or directions to objects from two or more points on the traverse line. In military surveys the lengths of the offsets are usually paced or estimated, and the bearings or directions are taken with a compass, or with the cavalry sketching board, when a plane-table is not used.

So far as military topography is concerned, traversing should never be employed when it can be avoided. The sketch of an area, for instance, is far better carried out on a plane-table or with a sketching case and prismatic compass triangulation than by a booked traverse.

Traversing can be used when no triangulation can be carried out, and it is thus useful in forests and in high grass. It may also be employed for the survey of rivers. When used in connection with a triangulation, or stations fixed by intersection, a traverse should always start and *close*, or end at a fixed point. When there are no fixed points a traverse should close on the starting point, or, if this be inconvenient, it should be checked by frequently intersecting some conspicuous object on the line of survey.

A traverse can be carried out in two ways:—

1. By recording all the observations in a *field book* and plotting them afterwards at leisure.
2. By plotting the observations as the survey proceeds.

Traversing with Compass and Field Book.

This method may be used—

1. When time does not permit the plotting of the work in the field.
2. When secrecy is desirable.
3. When the weather is too wet or stormy for plotting in the field.
4. When (as in the case of winding bush paths in W. Africa) the traverse is too complicated to plot in the field.
5. When it is necessary to keep pace with a rapidly moving column.
6. When high grass or forest prevent other methods.

The **Field Book** is a large pocket book with a column about three-quarters of an inch wide ruled lengthways down the centre of each page. This is sometimes called the *chain column*. The only entries made in it are the *forward bearings*, which give the directions of the traverse lines, the *distances* measured along those lines, and the *stations*. Of course any pocket book will do for a field book.

The chain column represents a line having no breadth. Therefore when a road or fence crosses the forward direction obliquely, it must be shown in the field book as arriving on one side of the column and as leaving it (Fig. 17) at a point exactly opposite on the other, and not as passing obliquely across the chain column.

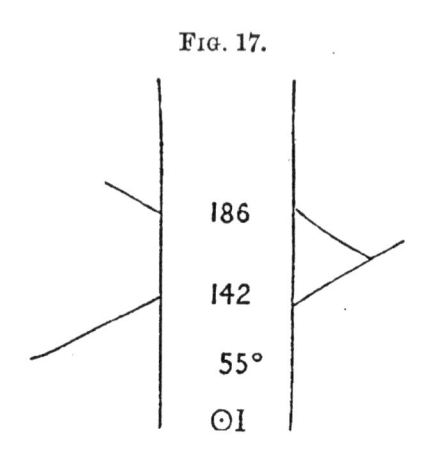

FIG. 17.

Forward bearings entered in the chain column must always be distinguished by the signs of degrees and minutes, so that they may not be mistaken for linear measurements.

The spaces on each side of the chain column are called the *offset columns*. All offsets are booked in these columns, and rough sketches of the detail, stream, fence, &c., to which they are taken are made in the same columns. Peculiarities of form in the detail are often exaggerated so as to convey more clearly the particular points to which the figures refer.

The length of an offset is entered to the right or left of the entry in the chain column that gives the distance on the traverse line at which the offset was taken. All offsets are taken at right angles to the traverse line.

All offset measurements are inclusive, that is, when two or three objects are on the same offset line, the distance of each from the traverse line is recorded.

Direction lines, showing the bearings to distant objects, should be drawn from the chain column.

The traverse entries should commence at the bottom of the page, and follow on in succession upwards.

For military purposes the only important traverses entered in a field book are those which record the details of a journey, and which will therefore be plotted on comparatively small scales. This is practically the only means of executing **sketches in dense forest**; the method is as follows —

SKETCHING IN DENSE FOREST.*

In many countries, parts of Africa especially, it is impossible, on account of dense bush, jungle, or grass, to survey any features other than the line of the road or track. As the paths wind about very much, the bearings have to be taken at short intervals, and such surveys being generally made whilst the sketcher is marching with a column or caravan, it is inconvenient and would occasion too great a loss of time to plot the bearings and distances on the march. It is preferable, therefore, to enter them in a note-book and to plot

* The method here described is also best for long continuous sketches from horse or camel in hot climates. If the bearings are long it is better to dismount from the horse when taking them (not necessarily from the camel). The plotting is done on arrival in camp.

them on arrival in camp. The bearings are better taken with a prismatic compass, if available, but if not, with an ordinary pocket compass. Distances are best arrived at by noting the times at which each bearing is taken or object passed, and estimating the rate of marching during the interval. A convenient form of notebook is formed by ruling four lines about half an inch apart down the centre of each page (see Plate XIV). Supposing the day's traverse is being commenced from Yandahu village, start by obtaining a back bearing to the last village. This is entered as B. B. (back bearing) to village 208°, and a forward bearing to the next village in front entered as F. B. village (forward bearing) Yolu 71°. In the thickest bush country a native has generally a very good idea of the relative positions of the villages round his own, and these bearings from village to village are very useful as checks on the direction of the traverse.

If a distant view is obtainable it is advisable to take back readings to any marked physical feature as far away as possible and near which the track has passed, also to any points in advance near which it appears probable it may pass. These long bearings of points, say 30 or 40 miles apart, are more useful than those from village to village in checking and correcting the direction of the traverse.

In the example (Plate XIV) bearings were obtained to two hills about 30 miles ahead. The bearing of the track immediately ahead is then taken and entered in the centre column, and the time in the first or left-hand column.

It is often impossible to see the general direction of the track, which probably disappears in the bush a few yards ahead; in this case it is necessary to have a native to show the direction, but constant watching is necessary, as many natives are very slow in understanding what is required, and seem unable to strike a mean between the direction of the path just a few yards in front and its position two or three miles ahead. To get a general direction of about half a mile of road, so as to take a bearing every 10 minutes, is a very convenient mean, and will give fair results. On arrival at the point where the direction changes and a new bearing has to be taken, the estimated rate of marching is entered in the third column.

If it be impossible, from a want of guides or other cause, to get satisfactorily the forward direction of the track, the best way of

PLATE XIV.　　　　　　To face p. 66.
SPECIMEN PAGE OF COMPASS TRAVERSE

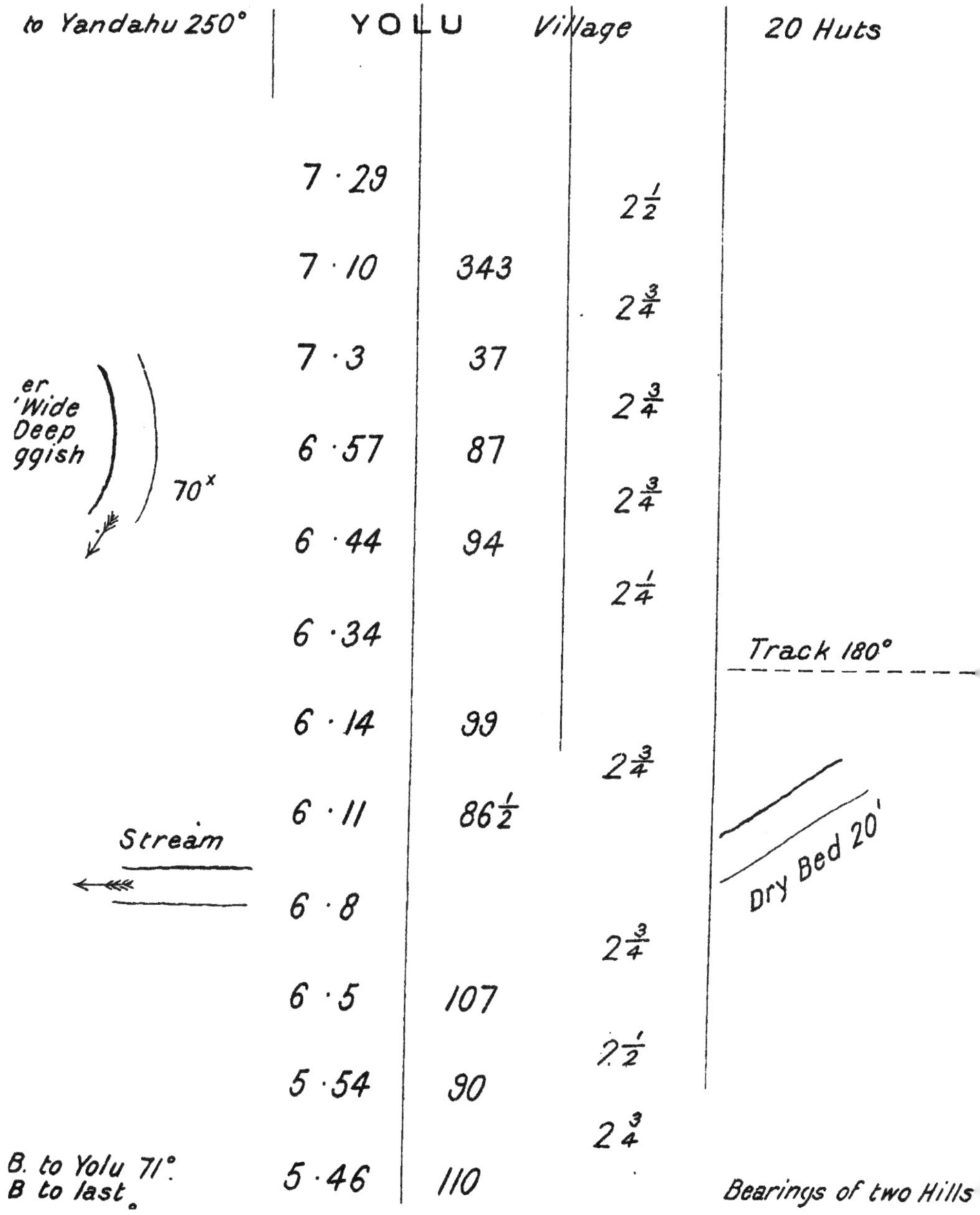

to Yandahu 250°	YOLU	Village	20 Huts
	7·29		
			$2\frac{1}{2}$
	7·10	343	
			$2\frac{3}{4}$
	7·3	37	
er 'Wide Deep ggish			$2\frac{3}{4}$
	6·57	87	
70×			$2\frac{3}{4}$
	6·44	94	
			$2\frac{1}{4}$
	6·34		
			Track 180°
	6·14	99	
			$2\frac{3}{4}$
	6·11	$86\frac{1}{2}$	
Stream			Dry Bed 20'
	6·8		
			$2\frac{3}{4}$
	6·5	107	
			$2\frac{1}{2}$
	5·54	90	
			$2\frac{3}{4}$
B. to Yolu 71°. B to last.	5·46	110	Bearings of two Hills

PLATE XV.
PLOTTING OF THE COMPASS TRAVERSE.
Shown on Plate XIV.

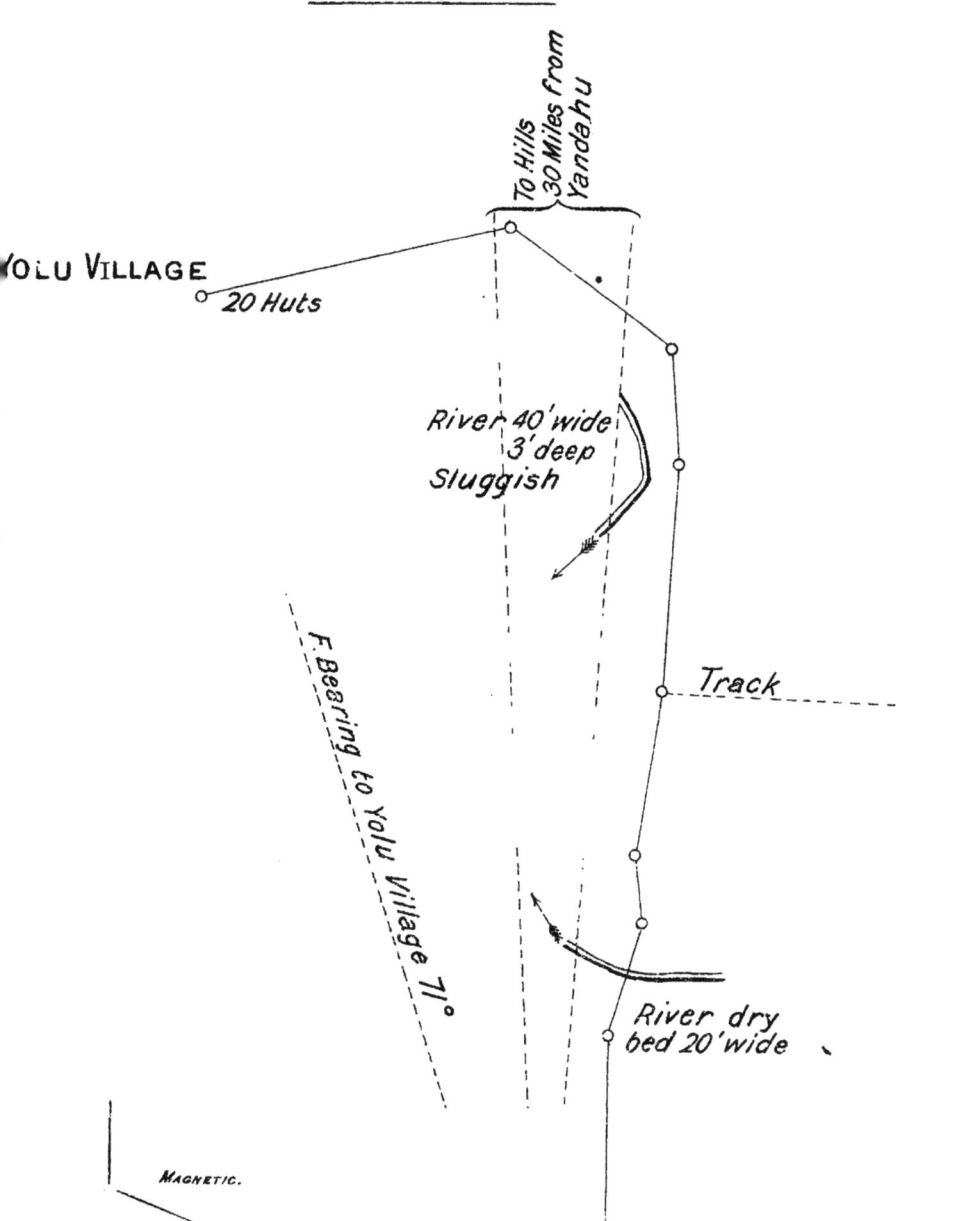

To face p. 66.

To Hills 30 Miles from Yandahu

YOLU VILLAGE
20 Huts

River 40' wide
3' deep
Sluggish

F. Bearing to Yolu Village 71°

Track

River dry bed 20' wide

MAGNETIC.

proceeding is to take the bearing upon which one happens to be marching at fixed intervals, say every five minutes of time, which may be shortened or lengthened according as the result is desired to be more or less accurate. Sometimes it is convenient to watch the compass swinging when held in the hand and take the approximate mean bearing as judged every two or three minutes.

It is necessary to watch carefully to see when the bearing of the path changes. If the sun be shining, one's shadow is a good guide; in the early morning, before the sun rises, it is very difficult to judge. In this case take a fresh bearing every five minutes.

The time occupied in taking the bearing and booking it as well as the time is so short as not to require any special consideration: these intervals can be allowed for in the estimation of the rate of marching. When, however, an appreciable delay is occasioned it is desirable to note it and make allowance for it in the plotting. In the example 6.14 is the time of arrival at the spot and 6.34 that of starting off again. It is sufficient to bracket the two times together to show what is intended.

The distances are plotted as straight lines: it is therefore necessary to allow in the "rate of marching" for the windings of the path. The maximum rate entered is $2\frac{3}{4}$ miles an hour, and this is as much as most caravans of porters will accomplish on an African bush track. If the track goes up and down hill, or is very stony or otherwise bad, it is necessary to reduce the rate entered accordingly.

After a little practice the error in length of traverses of this nature done with ordinary care should not exceed about eight miles in a hundred, and this is caused generally by a persistent over or under-estimation of the rate of marching. In direction the result is, as a rule, very good; it is wonderful how the number of small errors in bearings neutralise one another.

The track included in the specimen page, Plate XIV, is plotted on Plate XV.

For plotting it is convenient to make one or two scales showing the distances to be plotted in minutes of marching at the different rates.

Try to start and end up traverses of this nature on points of which the positions are known.

When marching with a column or caravan it is a good thing to work, if possible, for a mile or so beyond the camping ground at the end of the march, so as to be able in the morning to start off ahead and clear of the caravan.

It is advisable to carry the watch by some means which enables the time to be readily noted. A leather case, of the pattern used in South Africa, on the belt, or a band on the wrist, are both convenient methods.

It is a good thing to practice at first in a country of which there exists an accurate survey, as this gives a means of checking the work, and shows the topographer how much reliance he can place upon it.

Excellent traverses have been executed in a similar way during canoe voyages down rivers in Africa and elsewhere.

Traverses were made in S. Africa before the war by travelling in a waggon and noting the number of revolutions of the wheels, taking approximate forward bearings with a compass, and covering the margins of the pocket book freely with notes. Traverses of this kind may be of value.

Field book traverses on large scales are of little use for military purposes, and need not be practised by the student.

Traversing and plotting on the ground; is a method frequently employed when making a traverse for military purposes. A sketching board or case and protractor are taken, and each observation is plotted on the spot.

When working with a compass the paper should, before starting, be ruled with magnetic meridians so arranged as to keep the traverse on the paper.

On commencing work, select a convenient position on the paper for the first station, and mark it thus ⊙. Next observe and plot the forward bearing from this station and any bearings which may be taken to prominent objects, writing the name of each object clearly. Now commence pacing by marching on a point in the line of the forward bearing. On passing any important detail, such as a house, stream, &c., stop and, estimating the offsets, draw it in at once in its proper place. At the same time sketch in the details on either side of the traverse, and complete the plan with writing, &c., to the point reached. Now recommence pacing, and, taking up the count at the number at which the halt was made, continue the traverse in the same manner until a change of direction becomes necessary. Mark this point thus ⊙, take the requisite bearings, and commencing a fresh count of the paces, proceed as before.

Every opportunity should be taken of intersecting distant points, as they may be useful for checking the work or closing the traverse upon.

Attention should be paid to the scale of the sketch. The tendency is to show too much and to take too many bearings. This involves a waste of time, and the sketch becomes crowded with minute and unnecessary detail.

The sketch should be held in the hand (not attached to the body by a strap), and should be turned so that the direction of the protracted forward bearing may correspond with that of the same line on the ground. This enables the draughtsman to realise the connection between them.

All bearings should be lightly drawn, as they do not remain permanently on the sketch. Unnecessary prolongation of lines should be rubbed out, so as to avoid confusion.

All points at which forward angles are observed and a fresh counting of paces is commenced should be marked thus $\odot$, or otherwise distinguished in the sketch. The common error of laying off distances from the wrong point is thus avoided.

In open country the above method should only be used to connect points which have been fixed by compass intersections from a carefully paced base. In close country the method may be followed as above described.

In a civilized country very excellent traverse work may be done on a *bicycle* provided with a cyclometer; this does away with the most vexatious part of the old style of military sketching, viz.: the counting of paces, and is a rapid and accurate way of executing a military sketch. Time should not be wasted in pacing the length of houses and minor detail. As bicycles are usually made of steel, compass bearings should be taken at some little distance away from the bicycle.

CHAPTER X.

FIELD SKETCHING.

THE CAVALRY SKETCHING BOARD.

Sketching on Horseback.

The cavalry sketching board is a small board which can be strapped to the fore-arm. A magnetic compass is fixed in the board. Two rollers on opposite sides of the board enable a long strip of paper $7\frac{1}{8}$ inches wide to be carried, the paper being unwound from one roller and wound up on the other as the sketch proceeds.

The compass is fitted into a circular metal collar in which it revolves. A fine index line is marked across the glass, which can be adjusted to any required position by turning the glass. Before commencing to sketch it should be so adjusted that when the index line coincides with the needle, the direction of the paper is that of the general direction of the route to be followed.

The cavalry sketching board is thus used: The board is revolved until the magnetic needle in the compass coincides with the index line on the glass; when this occurs the board is "set." The directions of any required objects are then taken with a straight-edged ruler in much the same way as when sketching with the plane-table, the only difference being that no attempt is made to bring the eye down level with the ruler.

When sketching on horseback the following rules should be attended to:—

1. Turn the horse in the direction of the object.
2. Revolve the board until it is "set" as above.
3. By moving the left arm to the right or left, bring the point marking your position, and whence the line showing the new direction is to be drawn, *immediately under the eye.* Now align the ruler from this point in the required direction, and, placing the pencil point at it, draw in the line, *away*

from you. It is well to take a second look at the compass at the moment of drawing in the line, so as to guard against the board having been moved whilst the ruler was being aligned on the object.

Any slight alteration of the position of the board, in order to obtain a more exact coincidence of the needle and the index line, is best accomplished at this juncture, by gently turning the left wrist and forearm.

The correct alignment of the ruler on any distant point is a matter of judgment and "eye." With a little practice the ruler can be aligned and the direction drawn in correct to within 2° or 3° in between 10 and 20 seconds.

Most horses soon get accustomed to the work, and will stand steady enough for a brief period immediately after being reined up.

After a little practice the ruler can be dispensed with, except for long "shots," and the directions of objects drawn in by placing the pencil point at the place representing the position of the observer and drawing a line in the required direction.

It is clear that before starting work the compass should be "set," so that the direction of the road runs generally up the length of the paper. If in the course of the work the sketch runs off the paper, reset the compass and start afresh, having drawn a line right across the paper and marked the ending and starting point A in each position.

It should be fully recognized that accurate work is not expected with this instrument, but for executing a rapid route reconnaissance *on* horseback, where great accurracy is not required, its handiness makes it eminently useful; for this purpose the route should be simply "traversed" in the ordinary way, or in other words, a "Dead Reckoning" kept along the whole route.

The forward bearings should be taken to the furthest *visible* parts of the road, no matter how short or how long.

The clinometer is not a suitable instrument with which to keep up a *record* or *continuous* line of heights, so it is a necessity to have a pocket aneroid.

A reconnaissance of an area cannot be very effectively executed on a cavalry sketching board, as its drawing space is too limited to show any considerable area. For such a purpose a portable plane-table is advisable.

On horseback the cavalry sketching board should be strapped to the left arm, midway between elbow and wrist. The left hand (the bridle hand) should be kept high, or the pommel of the saddle will get knocked about by the edges of the sketching board. The ruler and pencil may be carried in the breast coat pocket, riding boot or legging, or in the hand.

As regards the measurement of distances when sketching on horseback, it used to be the custom to count the horse's paces. But this system has great disadvantages; it is vexatious and laborious, and the resulting preoccupation of mind prevents the sketcher from looking about him and studying the country through which he rides.

By far the best way is to measure distance by *time*.

Allow 12 miles an hour for a canter,
8 ,, ,, trot,
4 ,, ,, walk,

under normal conditions.*

Then draw a scale of minutes at the trot on the ruler. Thus :—
Scale of sketch 2 inches to 1 mile,—

In one minute at a trot $\frac{8}{60}$ mile, or 235 yards, will have been covered, 235 yards at 2 inches to 1 mile = ·27 inch.

In using this scale at the walk, halve it. Half time walking, half time trotting; take three-quarters.

Of course if it is known that the particular horse which is being ridden has not got the normal speeds, then make a special scale.

The scales to be used when making road sketches on horseback should be 1 inch or 2 inches to 1 mile.

The best pace to work at is a steady trot.

The principal use of a sketch executed as above is to serve as a diagram to illustrate a *road report* (see p. 96). The report being thus the essential part of the work, it will be seen that the sketcher must make notes as he rides along, and generally be on the alert. It is therefore *not* desirable to count paces. The report should be written at the bottom of the sketch; room should be left for it. Do not forget magnetic north point, scale, date, place, heading, and signature.

---

* Practice at keeping the horse at an even pace is essential.

CHAPTER XI.

FIELD SKETCHING WITH THE AID OF RANGE-FINDERS.

There is only one Service range-finder for all branches of the army, the mekometer.

The mekometer will often be found useful for executing rough sketches, especially sketches of country in occupation of the enemy.

The following is the method of using this instrument; for further information see "Handbook of the Mekometer."

TO TAKE THE RANGE.

THE two range-takers come up abreast, No. 1 with the reading instrument on the right, No. 2 with the right angle instrument on the left. The instruments are taken out of their cases, handles attached, telescopes focussed. The strap covers of the reels are unbottoned and tucked back under the belt out of the way. The instruments are held in the left hands.

The object, the range of which is required, having been pointed out to the range-takers, they consult together and determine on some particular part of the object which both shall lay on. They then each look outwards, as they stand facing the object, and select the direction in which to extend the base.

No. 1 turns to his right, takes a turn of No. 2's cord round his left hand; No. 2 turns to his left, and passes the end of his cord behind his back to No. 1 with his right hand.

(If a double base is to be used, No. 1 hooks the two cords together.)

Both numbers double out in the direction fixed upon, half the length of the base to be used (artillery, 50 or 100 yards; infantry, 25 or 50 yards), watching the object as they run so as not to lose its position. Arriving at the full length of the cord No. 1 halts, hooks his end of the cord on to the metal loop on the handle of the instrument, and faces the object, puts up the vane of his instrument, and revolves the drum to what he thinks is roughly the distance of the object. He plants himself firmly, his right foot somewhat to the front, and about 18 inches from the left (see Fig. 18). He puts up his instrument and remains steady. His left hand

FIG. 18.

Showing the two Observers in the act of taking a Range.

grasps the handle with three fingers, while the thumb and first finger steady the instrument (see Fig. 19). His right hand is on the drum ready to make his coincidence.

No. 2, on arriving at his end of the base, halts, hooks the end of the cord into the metal loop on the handle of his instrument, removes the telescope, and faces the object. He looks through his instrument and obtains his approximate right angle by moving backwards or forwards. If the reflection of No. 1's vane appears in the horizon glass to the right of the object he must retire, if to the left he must advance.

Short Rule.—Left—Advance, Right—Retire.

He then replaces the telescope in the instrument, looks through it, and effects coincidence, gradually stretching the cord taut as he does so.

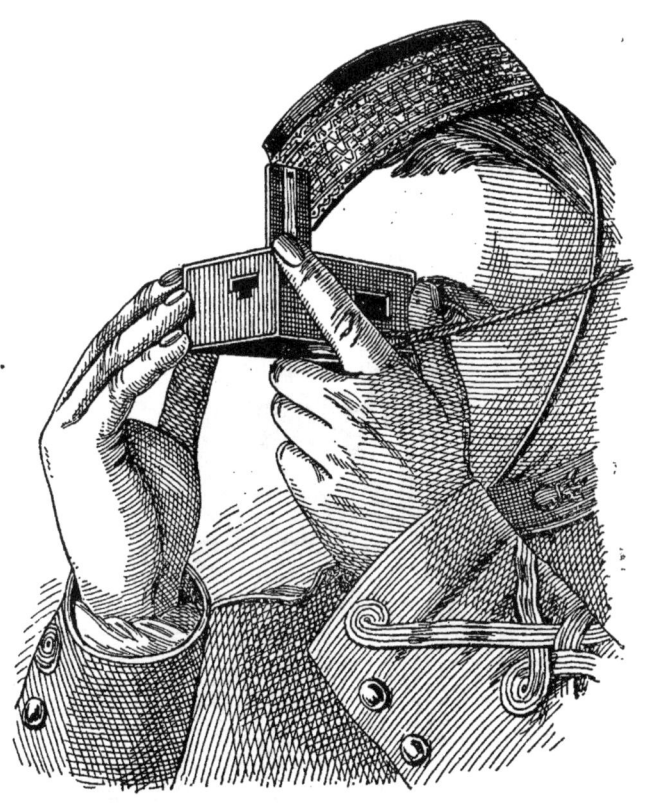

FIG. 19.

Showing the correct position of the hands in holding the Reading Instrument.

His position should be as follows;—Left foot well advanced, and slightly to the left, so as to get a steady pull on the cord (see Fig. 18). Left hand, two fingers round the handle; and thumb, first and second fingers holding three corners of the case; right hand, with fingers extended on either side of the cord, steadying the instrument (see Fig. 20).

He adjusts his right angle gradually by a slight movement of his body from the hips upward.

A correct coincidence, and therefore right angle, is obtained when the reflection of the white strip on No. 1's vane is seen overlying that part of the object previously agreed on between the range-takers. The coincidence should always be seen through the centre of the T-shaped opening.

Fig. 20.

Showing the correct position of the hands in holding the Right Angle Instrument.

Having obtained a coincidence he calls out "On," in a long-drawn singing tone, the object of this being to prevent the jerk, which would result from his calling sharply, making him lose his coincidence for a second or so. Having called he remains perfectly still, till he sees (by reflection) No. 1 take down his instrument to read the range. He can then rest his eye, but he must not move his feet or body, and must be prepared, the moment he sees No. 1 ready to take another observation, to at once get "On" again.

As soon as No. 2 has called "On," No. 1 turns his drum towards him till he has brought the reflection of the ivory strip on No. 2's vane slightly past, *i.e.*, to the left of the object. He then makes his final adjustment by turning the drum from him till the coincidence is complete, taking care to make it in the centre of the T-shaped opening.

It is an invariable rule that three readings should be taken before the range is reported. The drum should be turned off after each reading towards the operator, and the coincidence again adjusted by screwing up the drum. The three readings should not differ more than 2 per cent. from each other; if, therefore, No. 1 finds that one or more of his readings is too high or too low, he should call out to his No. 2 to give him more coincidences.

The above is the "drill" method of taking the range. Under Service conditions men must work as their intelligence prompts them, always bearing in mind the importance of keeping under cover.

Field Sketching with the Aid of a Range-finder.—A range-finder may be used in the execution of an eye-sketch, a compass sketch, or of a plane-table sketch; as the latter is more rapid and accurate, the method of its execution with the aid of a range-finder will be described. It will be assumed that it is required to sketch an enemy's position from a distance of 1 or 2 miles.

Set up the plane-table at a point giving a good view of the position. Draw the magnetic north line. Draw rays to all prominent objects in the position, marking houses, prominent trees, hill tops, junctions and bends of streams, bluffs and ridges, points on spurs, and any indications of the enemy's presence. Draw also a ray to the point where it is next intended to set up the plane-table —say a mile away. Take the ranges of all objects to which rays have been drawn.

Plot the distances along the rays, and sketch the country between the points so fixed, taking care to mark the points accurately fixed by a small circle. Move on to the next point and proceed as before, correcting any error in the sketching which is shown up from the new point of view

The method will be easily understood by studying the attached example (Plate XVI), which is a small portion of a plane-table sheet made in Natal on 5.12.99, by two officers and a small party of men. The instruments used were a plane-table and a Mekometer; the ground sketched was that north of the Tugela. The line traversed by the party, which was continually under fire, is shown in red; the stations from which ranges were taken are marked A, B, C; the distant points fixed are shown by small red circles. The scale of the map is 1,500 yards to 1 inch.

On completion of the map, copies were traced and sent with orders to the commanders of brigades.

This combination of plane-table and range-finder may be considered the practical and normal method of using range-finders for military sketching.

It is probable that as the accuracy and portability of range-finders increase, their use as an aid to field-sketching will become more frequent.

PORTION OF SKETCH OF TUGELA COUNTRY MADE (6 12 99) WITH PLANE-TABLE AND RANGE-FINDER DURING THE OPERATIONS

Scale 1500 yards to 1 inch.
1/54000.

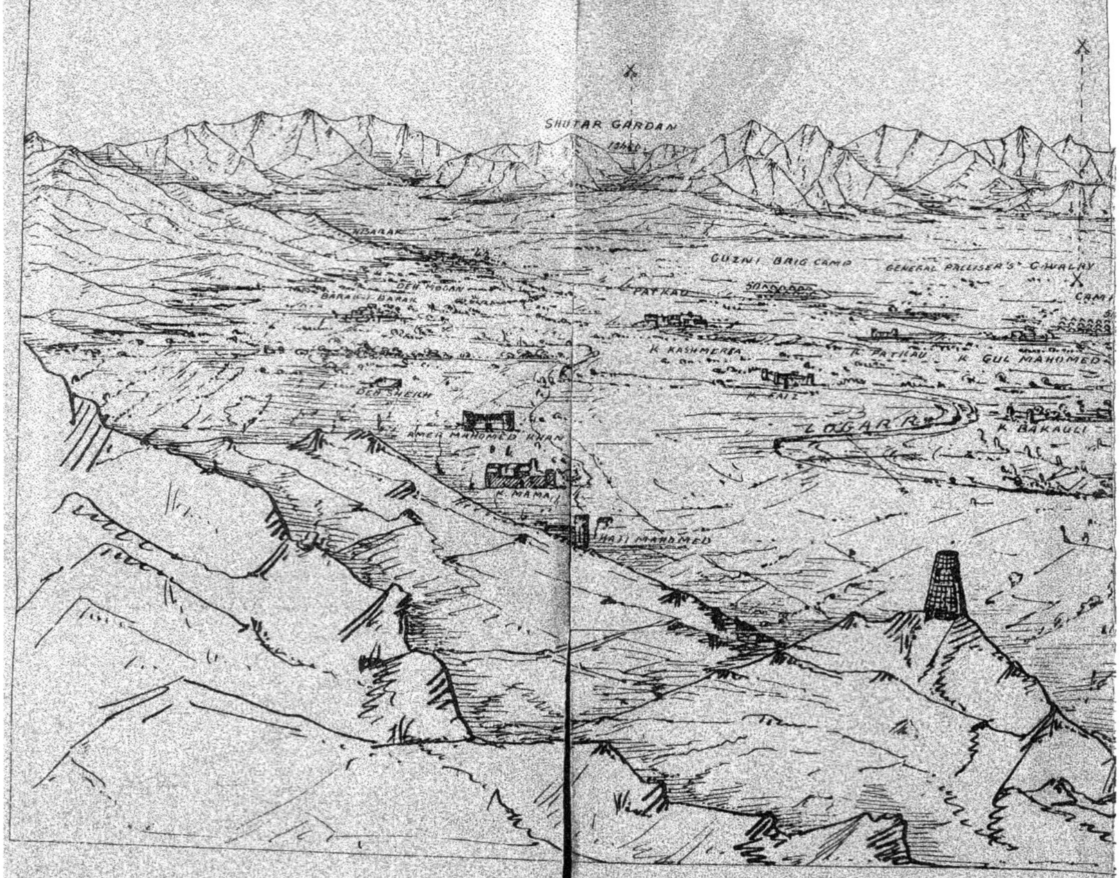
SHUTAR GARDAN
BARAK
GUZNI BAIG CAMP
GENERAL PALLISER'S CAVALRY
DEH MOGAN
BARAKI BARAK
PATKAU
K KASHMERSA
K PATKAU
K GUL MAHOMED
DEH SHEIKH
K FAIZ
LOGAR R.
K BAKAULI
AMER MAHOMED KHAN
K MAMA
HAJI MAHOMED

CHAPTER XII.

FIELD SKETCHING.

Landscape Sketching.

A LANDSCAPE sketch, especially of a position held by the enemy which cannot be closely approached, is sometimes of considerable value.

The sketch should be in outline, unshaded, except in the foreground.

The heading should state the position represented, the point from which it is sketched and the point of the compass which is faced (*i.e.*, which is opposite the centre of the sketch). Thus:—

Sketch of Linton Hill from Marden Railway Bridge facing north.

Names should be written opposite important features, and it is, of course, especially necessary to describe any indications of the enemy. Thus:—Enemy's trenches apparently from (A) to (B).

The ranges from point of sight of prominent objects depicted are also useful.

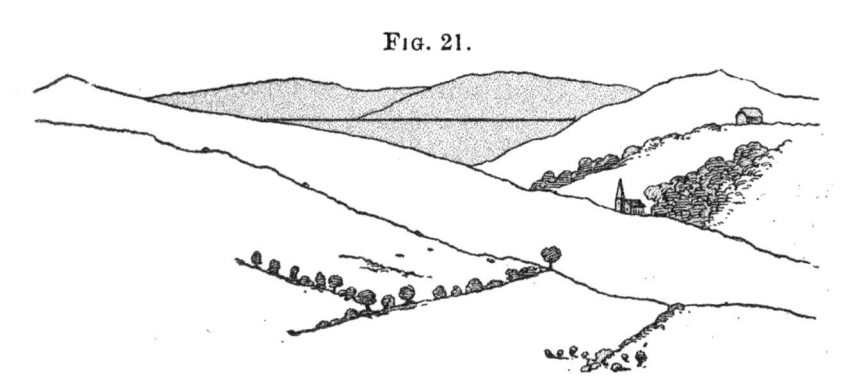

Fig. 21.

Such sketches are particularly easy of execution and useful when the features of the country are bold and well marked. It is permissible to exaggerate the heights of features which are important but do not show up well.

Examples of the use of landscape sketches are to be found in the Afghan War of 1880, the Chitral Campaign of 1895, and the South African War of 1899–1902. (See Plates XVII, XVIII.)

Even indifferent draughtsmen can produce landscape sketches which will serve to convey useful information to a General.

The leading idea of the draughtsman should be to show clearly the relative position and succession of ridges, valleys, and spurs, to mark on the sketch what villages, roads, tracks, or rivers are visible, and to write thereon such information regarding the enemy as he has been able to ascertain.

SKETCHING WITHOUT INSTRUMENTS.

Sometimes called Eye-Sketching.

It is assumed that a sketch is required and that no instruments can be obtained, but that the following are available:—(1) paper, (2) pencil, (3) a board to stretch paper upon, (4) some means of attaching paper to the board, (5) a straight-edged piece of wood.

First tie or pin the paper to the board. Then construct a scale to work with, thus: mark on the straight-edge a length which appears to be about $\frac{1}{2}$-inch thick, repeat this length along the straight-edge, by measuring along with a spare piece of paper. If the scale is to be 2 inches to 1 mile, each half inch will represent $\frac{1}{4}$ mile. Draw this scale also on the paper. If subsequently a protracter can be found, the true length of the scale can be measured and the representative fraction obtained in the usual way.

A good extemporised scale is a strip of ordinary ruled foolscap; the lines of this are as nearly as possible 600 yards apart on the 1 inch scale.

Now select a base, if possible with well defined ends. Lay the board on a bank, rock, or post at one end of it, or on the ground if a view can be got from it, mark a point to represent this end of the base and draw rays with the ruler to the other end of the base and to prominent objects within the area to be sketched.

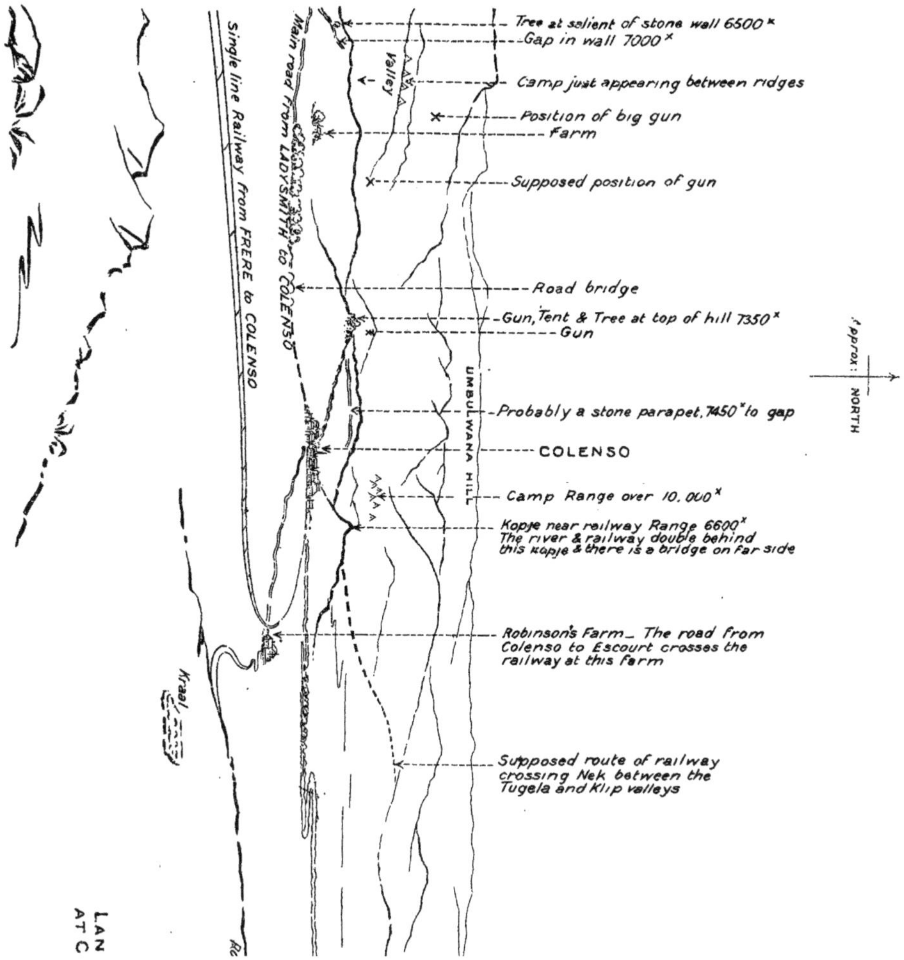

Tree at salient of stone wall 6500x
Gap in wall 7000x
Camp just appearing between ridges
Position of big gun
Farm
Supposed position of gun
Road bridge
Gun, Tent & Tree at top of hill 7350x
Gun
Probably a stone parapet, 7450x to gap
COLENSO
Camp Range over 10.000x
Kopje near railway Range 6600x
The river & railway double behind this kopje & there is a bridge on far side
Robinson's Farm_ The road from Colenso to Escourt crosses the railway at this farm
Supposed route of railway crossing Nek between the Tugela and Klip valleys
Single line Railway from FRERE to COLENSO
Main road from LADYSMITH to COLENSO
Valley
UMBULWANA HILL
Kraal
Approx: NORTH

Pace the base and mark off its length on the paper. At the other end of the base place the ruler along the line representing the base, and resting the board on some suitable object, turn it until the ruler is aligned on the first end, and then proceed as at the first end to draw rays to the same prominent objects and any new ones which may be visible.

A number of intersections will thus have been obtained. To obtain fresh intersections and extend the sketch, the observer should go to any point already fixed and set the board on one of the ends of the base.

The detail between these intersected points is to fixed partly by traversing (see p. 46) and partly by estimation. Never go over the same ground twice.

The principles to bear in mind in sketching without instruments are :—

1. Try to establish at the outset as long a base as possible, with well-defined objects at one or both ends, or in prolongation of the ends.
2. Always use the most distant points that have been fixed by intersection, in preference to the last traverse line, for " setting " your sketch.
3. In traversing, always try and work with the longest traverse lines possible, and use long " offsets," and do not change the direction of the traverse line to sketch in detail.

Fair accuracy is attainable when sketching without instruments, but the work is necessarily much rougher than when using a plane-table.

Every effort should be made to assimilate the conditions of work to those of plane-tabling. The absence of the tripod stand is sometimes much felt, since the board, when laid on the ground, is awkward to get at. Whenever a bank, fence, post, &c., is available it should be utilised as a make-shift stand to raise the board above the ground, and thus enable the rays to be drawn with greater rapidity and accuracy.

It is desirable that the approximate position of the north should be shown on the sketch. The true north can be laid down approximately by " setting " the board on the ground and drawing in a line which coincides with the shadow of a plumb-line at noon. In very cloudy weather when there is difficulty in noting the shadow at noon the rays to any distant conspicuous objects, such as mountain peaks, &c., should be drawn with the ruler, and names or description of the objects noted along the lines so drawn.

In any case the estimated position of the north should be laid down on the sketch, and the means by which it was fixed, noted against it.

For approximate methods of finding the north see Appendix.

CHAPTER XIII.

USE OF INSTRUMENTS BY NIGHT. MAP READING.

" Before* undertaking a night march, the roads or route must, as far as practicable, be carefully reconnoitred, and all points noted where checks are likely to occur.

" The general compass bearing should be previously noted and mentioned in the March Orders."

The officer who is to act as guide should be with the reconnoitring party, taking with him the best map of the district available.

In any ordinary country the advance by compass bearing only is attended by great difficulties, as such obstacles as woods, streams, and steep hill-sides necessitate any but a small force making a detour.

On a dark night in most countries it will only be possible to advance by well defined natural or artificial features, such as valleys, and streams or roads and railways, only marching on a compass bearing alone when absolutely compelled to.

The success of the operations will almost entirely depend on the completeness of the previous reconnaissance and preparations. During the reconnaissance the officer or other told off as guide should note what features, such as enumerated above, lead most directly in the direction he requires to go, making a special note of the best spots where it may be desirable to break away from one (say a road) to gain another (say a railway). These he should mark on his map. The compass bearing of the spot where he joins the new feature from the spot where he breaks away from the feature by which he has been previously marching should be noted. As few as possible of these deviations should be made, for it is much easier for a column to march by some well defined track than to follow a guide across an open plain : gaps in the column and consequent delays are almost certain to occur. When the recon-

* "Combined Training, 1902."

naissance has been pushed as far as possible, a compass bearing of the desired direction must be taken and recorded, and from this point the advance will have to be made by compass bearing only.

Marching on a Compass Bearing.

To use a compass with advantage on a dark night, it is essential that it should be prepared with "luminous paint." This substance must be kept exposed to the light before using it, or it will be found to be non-luminous when required. As a general rule, a compass exposed for half an hour before sunset will be sufficiently luminous to work by for some 6 to 9 hours afterwards.

The *Service prismatic compass* has been mentioned in Chapter VIII ; The magnetic north is marked by a broad black arrow-head. The compass is so constructed that upon the movable index being turned to point to the required "bearing," as shown on the external ring, and the arrow-head being made to correspond with the index, the *luminous line in the lid indicates the line of advance.*

It will thus be seen that in the service compass no attempt is made to read degrees or points by means of luminous paint : such a proceeding is impracticable, and if depended upon would only lead to disappointment and failure. The utmost that can be expected from luminous paint is for a compass prepared with it, to show a faint luminosity of the dial, just sufficient to throw up in relief the index and the black arrow-head. For experimental purposes luminous paint, immediately after having been exposed to daylight or magnesium light, will show up figures as clear as in daylight, but such conditions can rarely occur on service.

As a rule, the darker the night the greater the assistance afforded by the luminous paint. During a moonlight night a luminous compass is of no greater value than an ordinary compass with a clearly marked north point.

Luminous Maps.

When marching by night along roads in an enclosed country, it is a great advantage to have a "luminous map," which can be studied without the aid of a lantern. It is, however, obvious that these will not often be available.

The best and most durable maps for this purpose are drawn on service tracing cloth, and laid on a slab of luminous cardboard. To be read in this manner a map must be specially drawn as follows:—

The tracing cloth is laid on the map to be copied with the shiny side uppermost, and the road, &c., required, drawn with a good hard pencil. The cloth is then turned over and the pencil lines are inked in, with a broad pen, on the under surface, the lines being at least $\frac{1}{16}$ inch in thickness. The rules to bear in mind in drawing a map for use by night may be summarised thus:—

(1) Nothing should be shown that is not absolutely required.
(2) Any necessary printing should be in Roman block letters. $\frac{1}{4}$ inch in height, with $\frac{1}{4}$ inch clear space between them; when possible, initials or abbreviations should be used in place of whole words.
(3) No attempt should be made to show hills, &c., but any strongly defined line of heights may be shown by a thick line $\frac{1}{4}$ inch broad.
(4) The directions of the true and magnetic north should be shown at intervals along the route to be followed.
(5) Distances between recognisable points should be shown from time to time.

The most convenient way to use a luminous map of a route to be followed by night is to cut it into a long strip and mount it on a cavalry board. A piece of luminous cardboard about 5 inches by 8 inches can be slipped under the map to render the detail, &c., visible.

On a star-light night in the N. hemisphere such a map can be readily "set" at any moment by turning it until the true north point is directed on the Pole Star.

On dark nights, with no stars visible, the map can be set with the aid of a luminous compass.

In using a compass during a night march, some means should be adopted to prolong the line of advance as shown by the compass, and thus clearly indicate the route to be followed. A stick painted white, or with white paper pasted on it, or, best of all, one prepared with luminous paint, is most useful. The operator, standing perfectly steady, should wait until his compass has come to rest, and then hold the stick, at an angle of 45° to 60° with the horizontal plane, in the direction indicated by the luminous line in the compass-lid.

With the stick thus held it is easy to pick up some object to march on, and the advance is then resumed until it is considered desirable to halt, allow the compass to settle, and observe the direction of advance afresh.

The most favourable conditions for a night march are a clear starlight night and no moon. On such an occasion all that is requisite is for the guide to note the star which is on the alignment of his stick, as set by the compass, and then advance on that star for 10, 15, or 20 minutes, according to the position of the star in the heavens. He should then halt and take a fresh observation.

The most convenient stars to select for marching on are those which happen at the time to have an altitude of from $15°$ to $30°$; stars below $15°$ may become lost in haze, whilst for those above $30°$ the head has to be inconveniently raised. Speaking broadly, a star suitable for marching on will rarely move more than $5°$ in azimuth (*i.e.*, to a flank) in 20 minutes. Hence if at the end of every 15 or 20 minutes the observer halts and takes a fresh bearing, he can count on keeping direction correctly.

It should be clearly understood that although it is of assistance to an observer to know the stars, it is by no means *necessary* that he should do so. Anyone can, with care, march on a compass bearing, with the stars as points to direct him, with satisfactory results, although he may be totally ignorant of the names of the stars or even of the constellations. The great point is to be sure that the compass bearing is correctly taken and prolonged, and that the star selected is adhered to until a change is required.

But it may happen that, owing to clouds or fog, no stars are visible; in such a case, should there be no terrestrial objects to march on between the halts, the only safe way is for an assistant to stand behind the observer and work as follows :—The guide with the compass and luminous stick, as soon as the compass becomes steady, gives the word "steady." The assistant then, carefully noting the alignment of the luminous stick, advances in what he judges to be the right direction, until the guide halts him, before he is lost to sight, by giving a low whistle.

The guide having thus halted his assistant, notes by means of the compass whether the latter is standing on the true line of advance or to the right or left of it. He then moves up, and placing himself on what he judges to be the correct alignment, sends forward his assistant again.

The rate of advance obviously depends on the distances covered between halts; after the first few advances the assistant will have ascertained the number of yards he can safely proceed without being lost to sight, and on reaching that distance he will halt without waiting for the whistle. This is important, since the whistling might give notice to the enemy as the force approaches the objective.

Slabs of cardboard or wood, prepared with luminous paint, are of great assistance in this tedious process of sending on an assistant; a slab, 12 inches square, can be seen at from 70 to 150 yards distance, according to the condition it is in, and by this means the rate of advance can be greatly accelerated.

Besides the guide and his assistant, a third person should be employed to keep a careful record of the distances traversed.

Recapitulation.—Night marches, even in the most open country, should not (except in cases of absolute necessity), be undertaken without a previous reconnaissance. The march should be made where practicable by means of natural and artificial features, such as rivers and roads, and it is only in very rare circumstances that a march on a compass bearing would be made. If possible guides should be left at critical points during the reconnaissance. If a compass is used the Service compass is the best, and it may occasionally happen that it is possible to construct a luminous map, but it is impossible to count on having these ingenious arrangements on service, and the account of their use is only included in this book for the sake of completeness.

MAP READING AND USING MAPS ON THE GROUND.

Map reading is the term applied to the process of forming a clear mental picture of the natural and artificial features of that portion of the earth's surface which the map represents. The first thing to do on looking at a map is to note its *scale*. If the scale is expressed in terms of miles and inches it is easy to picture the real distances between the features represented. If the scale is expressed by a representative fraction only, it is sufficient to bear in mind that $\frac{1}{100000}$ is equivalent to a scale of 1·6 miles to inch nearly, and that $\frac{1}{250000}$ is almost exactly 4 miles to 1 inch, and other scales in proportion.

Next note the vertical interval of the *contours*, and see what are the heights of the hills above the lowest ground represented.

Determine whether the country is flat, undulating, hilly, or mountainous, or generally flat with occasional hills. Try to get an impression of the steepness of the hill sides by mentally comparing the country represented with country you are acquainted with.

Now observe the direction of the *north* (true or magnetic), the position of the *water-courses*, *streams* and *rivers*, and the general slope of the country, the position and direction of *ridges*, *watersheds*, and *spurs*.

Now note if the country is wooded or the reverse, the size of the villages and number of farms, the roads, railways, and canals, telegraph lines, bridges, and *artificial features* generally.

Making use of Maps on the Ground.

First study the map as described above.

Then, if possible, take it to some commanding point which can be identified; here it should be spread out and "set" so that its position may correspond with that of the country.

This may be done in three ways :—

(1) If any distant and conspicuous point is visible, turn the map until the line on the map joining your position, and that of the distant point, points to the latter.

(2) If any form of compass is available :—if the magnetic meridian is on the map, lay the compass over it and turn the map and compass until the latter reads magnetic north.

If only the true north is shown on the map, plot the magnetic north by laying off the variation (in England roughly $16\frac{1}{2}°$ to the W. of true north).

(3) The sides of most maps are approximately or exactly true north; hence a map can be set roughly by the sun; see Appendix V and VI.

NOTE.—The side margins of Ordnance Survey maps are nearly though not absolutely, true north; their greatest inclination in eastern and western counties is about 3°.

Or take the magnetic bearing of any conspicuous object, plot this line on the map from your position and see what point it passes through. Various other methods will readily suggest themselves.

If the observer is quite in ignorance of his position on the map, he can, if he can distinguish two distant known points, fix his position by taking their magnetic bearing and plotting back from them (resection).

Or, in the unlikely event of his being provided with a plane-table and sight-rule, he can fix the map on the plane-table and resect with great accuracy from three distant points, if recognizable, see p. 43.

Usually, however, the observer will have an idea as to where he is, and the exact position can be determined, first by "setting" approximately as above, and then by noting neighbouring detail.

Readiness in reading and using maps in the field results mainly from practice. It is important for officers to accustom themselves to read and use maps on those small scales (2 inches to 1 mile and under) which they are most likely to have to deal with on service.

"The attention of officers is drawn to the importance of training all ranks in the practical use of maps in the field, particularly of maps on small scales. It should be borne in mind that the employment of maps on large scales would be impossible on active service, and it is therefore undesirable that such should be used for peace practices or manœuvres ; . . ."*

* Regulations as to the Issue of Military Maps. 1902.

CHAPTER XIV.

RECONNAISSANCE.

Reconnaissance is the process of obtaining, recording, and supplying information concerning the movements, numbers, and dispositions of an enemy, or the nature and resources of a country, with a view to military operations. It is generally desirable that a reconnaissance should be undertaken for a definite purpose. The authority ordering a reconnaissance should issue instructions stating clearly the points on which information is specially required. The reconnoitrer should keep these instructions in mind, and should not overload his report with irrelevant detail.

It is not within the scope of this work to deal with the tactical handling of reconnoitring parties, which may vary in strength from a single officer up to a large force; for this, see "Combined Training, 1902." It is intended in this chapter merely to specify what information is required under various circumstances, and how it should be recorded.

The following account of **reconnaissances made during the Chitral campaign of 1895** will give an idea of the importance of the subject :—

The Chitral Relief Force consisted, at the beginning of the operations, of three mixed brigades, under Major-General Low. The object was to relieve the fort at Chitral, then besieged by the enemy. The route from India lay through Swat, Bajaur, and Dir. There were three or four recognised passes giving access to Swat, of which the Malakand and Shahkot were the best known, but no useful information regarding them was available. The country had not been properly mapped, and the detail on such maps as there were was very meagre.

During the march from the crossing of the Swat river to Bajaur an officer was detailed to make road reconnaissances in advance of the force. He was given, as a rule, a squadron of cavalry as escort.

The sketches, which were on a scale of 2 inches to 1 mile, were handed in each night to the General, who thus had the necessary information for the next day's march. No separate reports were made; notes were written on the sides of the sketches. The instrument used was the cavalry sketching board, and about 12 miles a day was sketched on an average.

This campaign also affords a very interesting example of the reconnaissance of a pass.

While the force was concentrating at Hoti Mardan, it became necessary to reconnoitre the Malakand and Shahkot passes, with the view of determining how best to cross into Swat. The same officer was detailed for this duty. It was not possible to go up to the passes, which were occupied by the enemy; but he got near the foot of the hills, and made free-hand sketches of them, as well as road sketches with the cavalry sketching board; this took two days. The result was to show that the passes were passable only by pack animals, and that with difficulty, and that the Malakand was the less difficult of the two. The G.O.C. therefore decided to cross by it.

A further reconnaissance of this pass was then made by another officer, who was given an infantry escort, and, under heavy fire, reconnoitred the ground up to the foot of the pass. The information obtained was particularly valuable, as it disclosed an unexpected valley, which had to be traversed before the pass was reached. It also showed that a spur closing in the near side of the valley afforded a good artillery position for the attack, and it was this position which was in fact occupied by the artillery, who fired over the heads of the infantry till they reached the top of the pass.

The above paragraphs refer to the reconnaissances carried out for the use of the leading brigade of the force. As the latter advanced, more detailed sketches and reports were made and sent to the troops following on in rear.

In general a reconnaissance consists of two parts :—

1. The Report.
2. The Sketch.

Framing Reports.

The following rules will be observed in drawing up a written report, in the field (Combined Training, Sec. 44) :—

i. The field message book (Army Book 153) and envelope (Army Form C 398) should be used, when obtainable. Officers commanding units will, on active service, procure these from Army Ordnance Department.*
ii. Clear writing is of first importance.
iii. Messages will be divided into paras., which will be numbered.
iv. The message must be as concise as possible, consistent with clearness.
v. The hour of 12 will be followed by "noon" or "midnight" written in words.
In describing a night, both days will be given thus : Night 29/30 Sept. ; or Night 30 Sept./1 Oct.
vi. Names of places and persons will be written in block capitals, *e.g.*, LONDON or WELLINGTON.
vii. If a map is referred to, the one used must be specified. The position of places will, as a rule, be denoted either by the points of the compass, *e.g,,* "wood, 600 yards S.E. of TETSWORTH," or by descriptions, *e.g,,* "point 276, close to the second E in HASELEY," the letter indicated being underlined. A road is best indicated by the names of two places on it. The terms "right" and "left" are used in describing river banks, it being assumed that the writer is looking down stream. Otherwise indefinite or ambiguous terms such as right, left, before, behind, beyond, front, rear, on this side of, &c., must not be used.
viii. If the message refers to troops reaching a place at a certain time, it is assumed that the head of the main body is meant, unless otherwise stated.
ix. The writer, having finished his message, should read it through carefully and ask himself whether it is clear, and, in the case of an order, whether it is calculated to influence the recipient in the way only that is intended.
x. The message must be clearly signed, the rank of the sender

* NOTE.—If, however, the report is illustrated by a sketch, as is usually the case in reconnaissances, the report should be on the same piece of paper as the sketch.

his appointment, and the force he is with, being stated It closes with the number of the message (all messages being numbered consecutively), place whence sent (fully described), date, official title (or name) of the person addressed, and the exact time of issue or despatch.

xi. A message sent simultaneously to several authorities will have the fact noted on each copy sent.

In writing a report it should be borne in mind that its sole object is to convey to another officer certain information. All verbiage and ceremonial style should be avoided.

Description of Country.—*Open country* implies ground free from obstacles, hedges, &c., which impede the view as well as the movement of troops.

Close or intersected country implies the reverse.

In some cases open country, where a good view is obtainable, may be intersected by dykes, and hence be impassable for troops; "clear view but intersected by dykes," would be applicable in such a case.

The nature of the ground, whether firm and passable by all arms, or boggy and passable only by infantry, or impassable to all arms, should also be reported.

Reports from natives should be taken for what they are worth. A note should be attached giving the sources of information, and an opinion should be expressed as to their reliability.

Reconnaissance Sketches.—A sketch is in many cases a necessary adjunct to a reconnaissance report, and almost always a useful one. It should only contain essential information. Clearness, not artistic effect, is what is required. As a general rule, the same information should not be given both in the report and on the sketch. In many instances the sketch and report may be combined, numerals on the sketch referring to notes in the margin.

The nature of the sketch will vary according to circumstances; in accuracy it may range from a rough diagram, serving to illustrate a report, to an approximately accurate survey.

A cutting from, or an enlargement of, an existing map will often take the place of a sketch. Care must be taken to check the accuracy of the map used. In many countries, *e.g.*, South Africa, owing to the absence of a topographical survey, the maps are most unreliable, whilst few maps of any country are quite up to date; new roads or railways may have to be added, or new plantations;

woods may have been cut down, villages and towns may have increased in size, windmills and other landmarks may have disappeared. It is sometimes a matter of importance to indicate the presence or absence of crops.

The following rules will be observed in making sketches:—
 i. Write at the top of the sketch the title or heading thus, "Sketch of Road from A to B."
 ii. Draw the magnetic north line at the side; also the true north line if the variation is known.
 iii. Draw the scale at the foot of the sketch.
 iv. State the vertical interval of the contours or form-lines.
 v. Use the approved conventional signs.
 vi. Give as much essential information on the sketch as is compatible with clearness. Refer by numbers to points mentioned in report.
 vii. Freehand sketches are sometimes of value even when drawn by an indifferent draughtsman. They must never be inserted for pictorial effect. In case of landscapes, ranges and distances should be indicated, and the point of view should be stated.
 viii. Sign and date.

Finishing Up.—Time is not to be wasted on this, but the sketch should be made clear for reference in a bad light. If possible, ink in; detail and lettering, black; water, blue; contours, brown. If ink is not available, use coloured chalks. If these are not to be had, finish in pencil.

Scales.—Road, river, and railway sketches would usually be drawn on a scale of 2 inches to 1 mile. The same scale is the most useful one for local purposes, *e.g.*, for the sketch of an outpost position, or of a defensive position. For general reconnaissances of an area, 2-inch, 1-inch, or ½-inch scales should be used. The only cases in which larger scales would be used are in sketching villages (for defence), sketching sites for camps, sketching *details* of railway lines and of outpost and defensive positions. In these cases, scales from 4 inches to a mile, and larger, may be used.

Training in Reconnaissance.

It is important that officers should in peace time accustom themselves when traversing a country to note its more important topographical features. The main points to note are:—

The slope of the ground and direction of watercourses.
Hill ranges and spurs; steepness of slopes.
Nature of ground, marsh, rock, plough, pasture, &c.
Woods and fences, and how they would affect movement of troops.
Roads, railways, canals, footpaths and tracks, rivers.
Villages and farms.
Positions for observation or defence.

Much benefit will be derived from the study of good maps in the field. For home service the hill-shaded 1-inch Ordnance map is a most valuable map; also the ½-inch Ordnance map.

A common fault is the employment of large-scale maps for such study. On the Continent the maps generally in use are rarely on a larger scale than $\frac{1}{80000}$, or about ¾ inch to a mile, and it may be taken that even smaller scale maps would more commonly be available. The disadvantages arising from the use of large-scale maps in peace time are that their absence on service would be felt, and further that, owing to the minuteness with which they show every minor feature of the ground, they induce a habit of constantly looking to the map in preference to the ground itself, when studying its tactical features. This habit should be striven against.

Readiness in estimating distances approximately is an accomplishment of great use on a reconnaissance. The best way to become a good judge of distance is to compare distances guessed at under various conditions of light and weather with the real distances as measured on a map.

On a rapid reconnaissance along road or river the most useful method of keeping a record of distances traversed is by time (see p. 65).

It will often happen that a range-finder is a valuable help in reconnoitring an area (see p. 73).

Sometimes distances may be roughly determined by measuring the time taken between the instants of seeing the flash and hearing the report of a gun. Sound may be taken to travel at about 350 yards a second.

The most important foreign measure of length is—

The kilometre (1,000 metres) = 0·6214 English mile, or 8 kilometres = 5 miles nearly.

DETAILS OF RECONNAISSANCE SKETCHES AND REPORTS.

A reconnaissance may be required for a great variety of objects. The following are examples. It may be assumed in the general case that both a sketch and a report are necessary.
1. **Road or Route Reconnaissance.**
2. **River Reconnaissance.**
3. **Railway   ,,**
4. **Coast   ,,**
5. **Reconnaissance of a Locality for Supplies and Accommodation.**
6.   ,,   ,,   **Woods.**
7.   ,,   **Villages and Farms.**
8.   ,,   **Defiles and Passes.**
9.   ,,   **Outpost Positions.**
10.   ,,   **a Position.** (*a*) for attack, (*b*) for defence.
11. ,,   ,,   **an Area** (topographical).

1. ROAD RECONNAISSANCE.

In a civilised country the old-fashioned road sketch, *i.e.*, a mere linear sketch between two points, is practically valueless. If the sole object of the reconnaissance is to ascertain the state of the communications, this is best done by sending out an officer provided with the Government map or an enlargement of it along the road or roads. Such remarks as are required being written in a separate report, which should refer to the map in question. Any point on the map not easily identified should be marked with a number.

A linear sketch may occasionally be useful in unmapped countries. But it is important to remember that generally any road sketch is of value mainly in proportion to the width of country sketched. And the term "Road sketch" is to be taken to mean not a mere sketch of a particular road, but a sketch of the country through which the road runs. Such a reconnaissance may often be valuable in discovering the reliability of the Government map in use. Maps are rarely up-to-date. Road reconnaissances in this sense, *i.e.*, sketches and reports of country through which certain communications run, should be carried out whenever a force halts for a day or so, and if made by a force on the march will sometimes be of great value to other bodies marching along the same route and to the Field Intelligence Division.

Road Sketch.—On foot, on bicycle, on horseback, cart or camel.

Scale usually 2 inches to 1 mile.

Interval of approximate contours 25 feet.

Commence at the bottom of the paper.

Usually drawn on cavalry sketching board; heights taken with aneroid barometer.

If the route runs off the paper, draw a line across and continue sketch in new direction up the centre of paper.

In the field a minimum of drawing is required; single lines for roads. Note width of metalled portion in feet along it, *e.g.*, 22′ m. for a road which has 22 feet of metal.

Show distances to important places off the sketch.

Dimensions of bridges, description of railways, note in margin.

Block in villages; show only principal buildings, such as church, town hall.

Show no fences.

Show direction of streams by arrows.

Write nature of country through which road runs thus: enclosed fields, marsh, open moor.

Sketch in woods.

Small free-hand sketches of prominent objects may be drawn in margin; these should not be attempted by those who cannot draw. Such sketches should be described, thus: "View of church A seen from (2), looking N.W., distance 2 miles." These are useful to identify turns and show at a glance which direction is to be taken.

In sketching a winding road, observe the direction as far as it can be seen, and draw in intermediate portions by eye.

In the case of a road reconnaissance the report is usually the essential part, the sketch serving mainly to illustrate the report.

It is not as a rule easy to find one's way by a road sketch; it becomes easier if as much as possible of the surrounding country is sketched.

Road Report* :—

i. *Roadway.* Metalled or not.

>Width in feet of metal.
>Level or hilly; give gradients.
>Good or bad condition.

ii. *Bridges.* Material; brick, stone, iron, or wood.

>Breadth and length.
>Number of arches; form, elliptical, segmental, semi-circular.
>Spans of arches or lengths of girders.
>Details of web and flanges of girders.
>Height of roadway above water or ground level.
>Measurements may be written thus:
>
>>10 yards—10$^{\times}$,
>>14 feet—14'.
>
>Nature of piers and abutments.

iii. *Towns and Villages.* Size and population.

>Material of houses, brick, stone, wood.
>Defensibility.
>Materials for barricading and entrenching.
>Entrenching tools available.
>Buildings available for barracks, hospitals, and magazines, such as church, town-hall, &c.
>Approximate number of troops which could be billetted.

iv. *Water.* Any near road fit for drinking or watering horses, stating numbers which could be watered at one time, and nature of approaches.

---

* NOTE.—This list of headings for the report and those of other classes of reconnaissance which are given in the following pages, are intended to draw attention to matters which *may* have to be reported upon. Only those headings should be used which are actually required. These lists, which are obviously highly academical, are not to be made the subjects of questions in examination, except when definite problems are set.

ky.

arms.

s and

to in-
1emy's

1ch as
passes,

ry re-
camp,
camp,

camp,

or $2\frac{2}{5}$

:annot

4.
ny of
mping
means

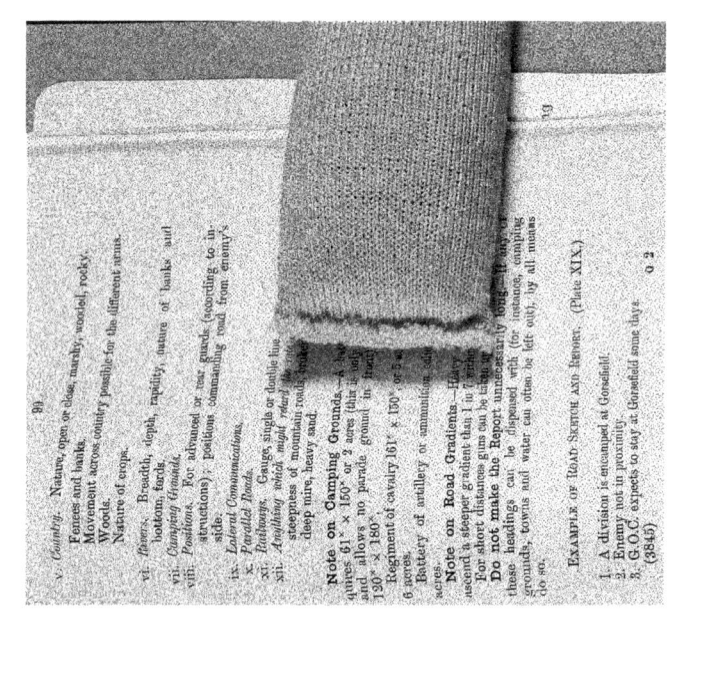

v. *Country.* Nature, open or close, marshy, wooded, rocky.
Fences and banks.
Movement across country possible for the different arms.
Woods.
Nature of crops.

vi. *Rivers.* Breadth, depth, rapidity, nature of banks and bottom, fords.

vii. *Camping Grounds.*

viii. *Positions.* For advanced or rear guards (according to instructions); positions commanding road from enemy's side.

ix. *Lateral Communications.*

x. *Parallel Roads.*

xi. *Railways.* Gauges single or double line

xii. *Anything which might retard* ...
steepness of mountain roads ...
deep mire, heavy sand ...

Note on Camping Grounds.—A bn... quires 61′ × 150′ or 2 acres (thirst ...) and allows no parade ground in fron... 120′ × 180′.

Regiment of cavalry 161′ × 150′ or 5 a...
6 acres.

Battery of artillery or ammunition co... acres.

Note on Road Gradients.—H... ascend a steeper gradient than 1 in 1 ...
For short distances guns can be tak... g

Do not make the Report unnecessarily long ... these headings can be dispensed with (for instance, camping grounds, towns and water can often be left out), by all means do so.

EXAMPLE OF ROAD SKETCH AND REPORT. (Plate XIX.)

1. A division is encamped at Gorsefield.
2. Enemy not in proximity.
3. G.O.C. expects to stay at Gorsefield some days.

(3845)

v. *Country.* Nature, open or close, marshy, wooded, rocky.
Fences and banks.
Movement across country possible for the different arms.
Woods.
Nature of crops.

vi. *Rivers.* Breadth, depth, rapidity, nature of banks and bottom, fords.

vii. *Camping Grounds.*

viii. *Positions.* For advanced or rear guards (according to instructions); positions commanding road from enemy's side.

ix. *Lateral Communications.*

x. *Parallel Roads.*

xi. *Railways.* Gauge, single or double line.

xii. *Anything which might retard the rate of marching*, such as steepness of mountain roads, broken places, rocky passes, deep mire, heavy sand.

Note on Camping Grounds.—A battalion of infantry requires $61^x \times 150^x$ or 2 acres (this is only for a temporary camp, and allows no parade ground in front). If standing camp, $120^x \times 180^x$.

Regiment of cavalry $161^x \times 150^x$, or 5 acres. If standing camp, 6 acres.

Battery of artillery or ammunition column, $75^x \times 150^x$, or $2\frac{2}{3}$ acres.

Note on Road Gradients.—Heavy service wagons cannot ascend a steeper gradient than 1 in 7 without extra horses.

For short distances guns can be taken up gradients of 1 in 4.

Do not make the Report unnecessarily long.—If any of these headings can be dispensed with (for instance, camping grounds, towns and water can often be left out), by all means do so.

EXAMPLE OF ROAD SKETCH AND REPORT. (Plate XIX.)

1. A division is encamped at Gorsefield.
2. Enemy not in proximity.
3. G.O.C. expects to stay at Gorsefield some days.

4. Meanwhile sends out officers to make reconnaissances in all directions within a radius of 12 miles, to ascertain—
- (a) Any news of enemy.
- (b) Reliability of ½-inch map in use.
- (c) State of roads, bridges, and fords, especially with reference to his steam transport.
- (d) State of telegraphs.
- (e) Supplies (reported on separately).

1b. ROUTE RECONNAISSANCE.

This is only a road reconnaissance of an extended nature and on a smaller scale, say ½" or ¼" to 1 mile. The system on which the sketch is executed is described on p. 65. If a report is required it should be tabulated as a road report.

2. RIVER RECONNAISSANCE.

There are three cases in which a reconnaissance of a river may be required:—
1. When the near bank only can be traversed, the opposite bank being in possession of the enemy (ex. Tugela, 1899).
2. When the stream and both banks are accessible to the reconnoitring officer.
3. When the enemy is in possession of the surrounding country, and the only possible reconnaissance must be made by water (ex. Nile Expedition, 1885).

The *sketch* would usually be on a scale of 2 inches to 1 mile.

The following are headings for the *report* which may be required; an officer making the reconnaissance should only use those which are wanted —

i. *The valley.* Nature, whether swampy, rocky, wooded.

Give all information possible on this head on *sketch*.
The communications through it.

ii. *The stream.* If navigable or liable to floods.

Breadth, depth, nature of bottom, rate of current.
Channel, breadth, and depth.
If tidal, rise and fall.

iii. *The banks.* Steep, shelving, marshy, or firm; height, stating if there is any marked difference in level between them; whether roads or tow-paths run along them; cover for firing line; is movement along them and towards them free or impeded by ditches?

iv. *Tributaries.* Full details; note especially whether sufficiently navigable for collecting, joining, and floating down bridging materials under cover from view or fire.

v. *Islands.* Connected by fords? Will they help to bridge river?

vi. *Bridges.* If permanent, as in road report.

If floating, size and number of boats, pontoons, and whether passable by artillery.

vii. *Fords.* Position, nature of bottom, length, breadth, and depth.

Defensible or reverse.

viii. *Points suitable for bridging.*

ix. *Approaches.* Nature of roads and tracks.

Largest front on which three arms could move.
Can they be improved?
Places where approaches could be barred by obstacles or held by troops.

x. *Boats, flying bridges, and ferries.* Make classified list of boats obtainable, giving approximate dimensions and number of men each would carry.

xi. *Locks* and weirs.

xii. *Inundations.* Means of effecting, if required.

xiii. *General tactical considerations.*

(a) *Advance.* Points favourable for forcing a passage.

Artillery positions for covering same.
Enemy's positions; can they be commanded or enfiladed?
Has he good artillery positions?
Villages, woods, or broken ground under cover of which troops could be massed.
Can troops be sent down to river unseen by enemy?
Can defensive positions on far bank be seized and entrenched to ensure passage of main body?

(*b*) *Retreat*. In case of retirement, defensive positions to prevent enemy getting within range of passages by which force retires. Arrangements for demolishing bridges after retirement.

Defensible positions on near side to hinder enemy from debouching after crossing.

(*c*) *Opposing hostile advance*. Can our troops be readily brought up to threatened point, and what defensive positions can be taken up to cover it ?

Do not write a long-winded report. Report only on those points which affect the object of the reconnaissance.

Note on Velocity of Rivers.—The strongest current is usually in the centre of straight reaches, and near the concave bank in curves.

The following expressions are sometimes used to describe velocities :—

Sluggish, not more than $1\frac{1}{2}$ feet a second, or 1 mile an hour.
Ordinary,   ,,   3   ,,   2 miles ,,
Rapid,   ,,   5   ,,   3   ,,   ,,
Very rapid,   ,,   8   ,,   5   ,,   ,,
A torrent, anything faster than the last.

A fall of 6 inches in a mile will produce a current of 3 miles an hour in a large and deep river.

Note on Fords.—The most certain way to search for fords is to drop down stream and sound with a lead attached to a line of the required length, or with a pole. Having located a ford, it should be examined by crossing it several times. As a rule, the reports of inhabitants as to the existence of fords, even in a friendly country, are unreliable; it commonly happens that excellent fords are to be found of which the inhabitants know nothing, either from the river having silted up without their knowledge, or from local custom having adopted certain places to ford a river which happened to present some advantage, real or imaginary. Fords are generally to be looked for above or below sharp bends, and they almost always run diagonally across a river. Sometimes a good ford will lead to a shallow down the centre of a stream, which may have to be followed for a considerable distance before it is possible to cross from it to the far bank.

Fords of this nature should be carefully marked by stakes when possible, and if shown on a small-scale reconnaissance sketch, should be further illustrated by a hand-sketch of the ford in the margin on a larger scale, showing exactly how to cross the river at such a point.

The depths usually considered passable in fords are as follows:—

	Rapid stream.	Sluggish stream.
Cavalry	48 inches	54 inches.
Infantry	36 ,,	42 ,,
Guns and wagons	30 ,,	30 ,,

Frozen Rivers.—Should a river freeze, it is useful to remember that ice 3 inches in thickness will bear infantry. and from 4 to 7 inches, cavalry and field guns. The thickness of ice can be easily increased during a frost by covering it with rushes or straw and pouring water over them.

The width of a river can be determined—

1. By a range-finder.
2. By plane-table.
3. By observing compass bearings and plotting to scale.
4. By elementary methods (" ground problems ").

Reconnaissance down a River.—In those cases, which are sufficiently common in tropical Africa, when it is required to make a reconnaissance while travelling down a river in a canoe, boat, or launch, all that can be done is to make a timed compass traverse as described on p. 65. If in a launch which has a binnacle, of course this would be used instead of the prismatic compass.

3. RAILWAY RECONNAISSANCE.

A railway reconnaissance will be usually made with a view to ascertaining—

(1) Its capabilities for the transport of troops.
(2) The best means for its destruction or repair.
(3) The nature of the neighbouring country as affecting military operations.
(4) Its capabilities for use as a road along which to march troops (as in the American civil war).

It is difficult to imagine circumstances under which an expert engineer officer would not be available for duties (1) and (2); but if other officers have to undertake it, the following brief notes will serve as a guide. The reconnaissance is best made on a locomotive going about 4 miles an hour, or on a trolley.

The *sketch* may be on a scale of 2 inches to the mile. Show each pair of rails by a single red line; exaggerate the widths so as to show sidings and cross-over roads in more detail. Large scale sketches of stations may be wanted.

The *report* should contain the following headings:—

  i. *Construction of Line.*—Gauge from *inside to inside* of rails; rail used, how fastened; sleepers, iron or wood; chairs or spikes; how ballasted; general state of repair; maximum gradients and curves; cuttings, embankments, tunnels; distance between stations or signalling stations; junction lines; signal boxes, and method of working same.

  ii. *Bridges.*—Length, spans, width, materials.

  iii. *Stations.*—Length and breadth and position of platforms, height above rails; facilities for lengthening or making platforms, sidings, and end loading docks; means of entraining men, horses, guns, wagons; means of watering engines, fuel for engines; drinking water; entrances and approaches; shunting arrangements, turntables, cranes, and engine sheds.

  iv. *Rolling Stock.*—Amount and description, carriages of each class, cattle and goods trucks, number and description of engines. Repairing facilities. Workshops.

  v. *Stores.*—Spare sleepers, rails, chairs, spikes, fish-plates, bolts, telegraph material, tools.

  vi *Coal Depôts.*

  vii. *Personnel.*—Officials, engine drivers, pointsmen, gangers, minimum number of men required for signalling arrangements.

  viii. *Telegraphs.*—Number of wires; telegraphic apparatus.

  ix. *Demolitions.*—Best method of rendering line unserviceable, most favourable points for removing rails (on a curve remove *outer* rail); bridges and tunnels to destroy.

  x. *Defence* of tunnels, bridges, culverts and stations, and the line generally.

4. COAST RECONNAISSANCE.

The object of this is to obtain information as to the suitability of the coast—

(a) For embarkation or
(b) „ disembarkation, or
(c) „ observing and opposing a landing.

It may safely be assumed that such a reconnaissance would generally be carried out by the navy, or in conjunction with the navy. The essential hydrographic information would usually be taken from the Admiralty charts and sailing directions.

The points to be reported upon are:—

i. *Anchorage*, roadsteads, and shelter from wind for transports.
ii. *Proximity of 5-fathom line* to coast; nature of coast, whether shelving or not.
iii. *Tides and currents;* rise and fall of tides, hours of high and low water.
iv. *Facilities for landing;* foreshore, sand, rock, or shingle; points for construction of wharves and piers.
v. *Existence in the vicinity of any harbour* or estuary which would serve as one which might be taken possession of to use as a base.
vi. *Existence of any defensive works.*
vii. *Positions* which might be occupied by covering party.
viii. *Communications* on shore.
ix. *Water, food, fuel* on shore.
x. *Country in vicinity*, whether suitable for movement of troops. Tactical considerations.
xi. The prevalence or otherwise of surf in windy weather.
xii. How near can war vessels approach the shore to cover the first disembarkation.

5. RECONNAISSANCE OF A LOCALITY FOR SUPPLIES AND ACCOMMODATION.

When statistical information is required about supplies and accommodation along any road or route, it is better to embody it in a separate tabulated report. Where possible, the local Directory should be secured. Every officer reconnoitring in the field should accustom himself to note what forage, cattle, sheep, or other supplies are to be met with in the tract of country he is examining.

In *quartering troops* on the inhabitants in civilised countries, it is necessary to decide rapidly on the number of men that can be accommodated in any house. The following rule will be found sufficiently accurate:—

For every room 15 feet wide or under, allow one man to every yard of length.

For every room over 15 feet, but under 25 feet, allow two men for every yard of length.

For rooms 25 feet wide, allow three men for every yard of length.

Frequently it will be quite impossible for an officer engaged in quartering troops to visit more than a few houses. In such a case the better plan is to divide the houses into classes, and by examining a house of each class, decide upon the number of men it will accommodate. Some rooms, usually those on the top floor, should be left for the inhabitants.

In all cases, extra space must be allotted to troops which are in permanent occupation of houses. In an emergency, and for a short time, men can be crowded in to obtain shelter from the weather. The minimum space for a soldier and his accoutrements is 3 feet in width by 8 feet in length, and when there are several rows of men sleeping on the floor, passages of at least 1 foot wide must be left between each row to enable men to pass in and out of the room.

In reckoning accommodation for horses in barns or outhouses, about 5 feet of their length should be allowed for each horse. If a building is 24 feet wide, two rows of horses can be placed in it.

Under the head of "*Supplies*" are given the names of the principal persons upon whom requisitions for flour, corn, hay, oats, and live stock might be made, stating the amount obtainable from each person, butcher's, baker's, and blacksmith's shops; fuel and corn stores, hotels and inns, forges, mills, and wheelwrights.

For calculating the supply of fodder:—Hay weighs 200 lbs. per cubic yard in the rick; straw 140 lbs.; grain 900 to 1,300 lbs. per cubic yard; oats are the lightest.

Under the head of "*Transport*" should be given the names of those persons who possess carts or wagons adapted to the transport of commissariat and other stores, with the number of horses belonging to them. Also the names of persons on whom requisition can be made for saddle horses.

Under the head of "*Water.*" Whence is the water supply available? Is the supply limited? Is the water of good quality?

To find the supply of cubic feet of running water per minute. Calculate the area of the cross section of the stream in feet (breadth by average depth); multiply by velocity in feet per minute. The velocity is easiest found by timing a piece of wood thrown into mid-stream. The mean velocity = $\frac{4}{5}$ surface velocity.

Having thus ascertained the capacity in cubic feet of the water supply per minute, multiply it by 6·23, and the result will give the number of gallons per minute.

An alternative method for obtaining the water supply of a stream in 24 hours is as follows:—Area of the cross section (in sq. feet) multiplied by the velocity (in feet) per minute, multiplied by 9,000. (The number 9,000 is obtained by multiplying the approximate number of gallons in a cubic foot of water ($6\frac{1}{4}$) by the number of minutes in 24 hours.)

About 1 gallon per man per day for drinking and cooking is sufficient; 2 to 3 gallons are required for washing. In stationary camps 5 gallons per diem should be allowed per man. About 10 gallons should be allowed for each horse.

To calculate the amount of forage rations in a rick, measure the rick in yards and ascertain contents by the following formula:—

Contents = length × breadth × (height to eaves + $\frac{1}{2}$ of height of roof).

If the rick be of hay, divide this amount by 12 to obtain the number of tons it contains. If it be of straw, divide by 17.

6. RECONNAISSANCE OF WOODS.

This would usually be an item in a larger reconnaissance, as of a position. The points to note are:—

i. Position and extent.
ii. Communications.
iii. Nature of trees and undergrowth, passable for troops or the reverse; suitability of timber for constructing abattis.
iv. Nature of ground, dry, marshy, level, or hilly, streams or other obstacles.
v. Second or third line. Existing clearings. Any buildings with clear space round them.
vi. Whether exposed to artillery fire or not.

7. RECONNAISSANCE OF VILLAGES OR FARMS.

Villages, farms, &c., are often of considerable importance for local defence. For example, a village is commonly found adjacent to a bridge over a large river, and so forms a part of the defensive system should it be required to hold the bridge. As strong points on a field of battle they are constantly met with.

In reporting on a village, &c., for defensive purposes, the following would be useful points to examine:—

i. *The Village.*—Situation, whether exposed to artillery fire, and at what range; construction, whether inflammable, how surrounded and enclosed, whether a strong shooting line could be established along its exterior line, and whether interior arrangements suited for defence.

ii. *The Surrounding Ground.*—Whether open and affording a good field of fire, or enclosed and offering good cover to assailants.

iii. *Interior Defences.*—Whether a second and third line of defence could be created, buildings suitable as keeps

iv. *Communications.*—Whether open spaces for reserves, &c., exist, nature of interior communications, how to improve the same, also facilities for dividing the village into "sectors" so as to localise any success on the part of the assailants should they gain a footing.

v. *Position of Reserves.*—Places suitable for posting strong reserves under cover in rear or on the flanks of the place, and moving them up for counter-attack.

vi. *General Considerations.*—Best place for garrison during a cannonade, suitable positions for artillery of defence in vicinity, strength of garrison suited for place.

vii. *Proposed Strengthening.*—Measures most suitable for strengthening lines of defence, loopholing, trench work, &c., barricades necessary.

viii. *Demolitions and Obstacles.*—To remove any outlying cover which might shelter an assailant, and to obstruct any advance where the defence was weak.

A table of work proposed, showing places, number of men, time required, tools required, &c., should be appended to the report.

A sketch of a defensive village, &c., would usually be on a much larger scale than an ordinary reconnaissance sketch; 8 inches to 12 inches to a mile would be a suitable scale for showing any of the preceding arrangements.

Hand-sketches showing sections of any line of defence, &c., should be added when necessary.

8. RECONNAISSANCE OF DEFILES AND PASSES.

Any natural feature which causes a force to contract its normal front during its passage through it, is a *defile* for that force. A mountain pass is the most common sort of defile. A good map, even if on a small scale, is of the greatest value.

As in other reconnaissances with a view to fighting, frequently the best idea of the capabilities of a mountain defile for attack or defence will be obtained from the side of the enemy during the return journey; for on ground of a broken nature it is often difficult to judge of its exact strength or weakness when viewed from the side of the reconnoitrer. Hence any points about which notes were made during the advance should be reconsidered, and, if necessary, added to or corrected during the return march.

The *sketch* should usually be on a scale of 2 inches to 1 mile.

The *report* should be generally the same as any other road report, with special attention to the following points:—

 i. *The Defile.*—Length, breadth, open spaces, view obtainable at various points, whether exposed to fire throughout.
 ii. *The Flanks* of the enemy's position; can they be turned?
 iii. *The Route.*—Nature of track, gradients, roads or tracks which join it.
 iv. *Country beyond.*—Exit commanded by artillery positions? room for debouching; positions which might be occupied.
 v. *Positions.*—Strong points which would serve for local defence; positions which might be held by the enemy; artillery positions.
 vi. *Obstacles.*

In this sort of reconnaissance landscape sketches may be most useful.

See pp. 79, 91, and Plates XVII and XVIII.

9a. RECONNAISSANCE OF OUTPOST POSITIONS.

An officer charged with the duty of reconnoitring ground for an outpost position would, as a rule, be given the general line which the main body is to occupy.

"An outpost position, so far as possible, should possess the following characteristics:—

i. It should be strong for defence.
ii. It should be difficult to surprise.
iii. Retirement from it should be easy.
iv. Command, co-operation, and intercommunication will be facilitated by placing the advanced groups along well-defined natural features, such as ridges, streams, the outer edges of woods, &c., or in the vicinity of roads.
v. If the outpost position includes commanding ground, from which a wide extent of country can, in clear weather, be kept under observation by day, it will be a great advantage. Facilities for observation, however, are of less importance than facilities for protracted resistance. . . . . Commanding ground is advantageous, but by no means essential." ("Combined Training," 1902.)
vi. The flanks of an outpost position will be well secured if they rest on some natural obstacle, such as a river or morass.
vii. It should deny the enemy positions from which they might shell the camps.

The *Report* should describe the general line of front for the sentries to take up, the positions of piquets, supports and reserves, detail the day and night arrangements respectively, also what scouting should be done by cavalry, the most suitable line of resistance, points for signal stations, water supply, communications, general nature of country.

The *Sketch*, usually on the scale of 2 inches to 1 mile, should show the general arrangements. An enlargement from a small scale map will do, or a tracing of such a map, numbered or lettered so as to show the position of the various portions of the outpost line.

The details of an outpost position should always be settled *on the ground*, never on the map. The map or sketch is to show the general organisation of the position.

9b. RECONNAISSANCE SKETCH BY AN OFFICER ON OUTPOST DUTY.

Every officer in command of a piquet will send a rough sketch of the post and ground in its vicinity to the commander of the outpost company to which he belongs. It may be conveniently done in a pocket book on a scale of 4 inches to 1 mile. The sketch should show :—

(a) The position of the piquet and the point it would occupy if attacked.
(b) The position of sentries by day and night.
(c) The direction and distance of the piquets on either flank and of the support and reserve by means of arrows, thus:
⟶ To No. 2 Piquet, 800 yards.
(d) The *ranges* of any woods, farmsteads, or other localities which would probably be occupied by an enemy advancing to the attack or near which he must pass.
(e) The most favourable line of retreat.
(f) Direction and distance to which patrols are sent.

10. RECONNAISSANCE OF A POSITION.

The principles which determine the selection of a position for defence are described in "Combined Training," 1902. The chief requisites are:—

i. The locality should satisfy the strategical object.
ii. The extent should be suitable to the strength of the defending force. As a very rough rule two battalions (at full war establishment) entrenched should be sufficient to occupy a mile of front, exclusive of the general reserve.
iii. A clear field of fire to the front.
iv. Flanks to rest on ground naturally strong or made so.
v. Good cover.
vi. Good artillery positions.
vii. Sufficient depth; good lateral communications.
viii. Good means of retreat.
ix. No good positions for enemy's artillery.
x. Favourable ground for counter-attack.
xi. Water.

NOTE.—Trenches or guns on the sky-line afford so excellent a target that such a position, especially if the enemy has good artillery, should always be avoided.

Trenches should always be concealed by all means possible.

(*a*) *For Defence.*—The *Report* should deal with the following points:—

1. Length of position and numbers for which it is suitable.
2. Nature of the flanks.
3. Nature of surrounding country, whether close or open, favourable or otherwise to movement by the three arms, flat or hilly, ground hard or swampy, what obstacles intersect it, villages, woods, &c., and their nature.
4. Interior communications.
5. Approaches to the position from the enemy's side.
6. Lines of retreat.
7. Suitability or otherwise for development of artillery fire.
8. What hostile artillery positions exist, and their ranges.
9. Facilities for counter-attack.
10. Whether advanced posts necessary, and why.
11. Depth of position and capabilities for retrenchment.
12. Facilities for construction of obstacles.
13. Nature of soil as affecting entrenchments.
14. Suitable sites for entrenchments.
15. In certain cases what supply of civilian labour, tools, and transport is available.
16. Water supply.

A position should always, where possible, be looked at from the side from which it is exposed to attack; any weak spot should be brought to notice.

The Sketch.—When a topographical map is available, use it, or make an enlargement from it; otherwise the scale will depend on the extent of country. The special maps made for defence purposes of Bloemfontein and Pretoria were on a scale of 2 inches to 1 mile. Put as much on the sketch and as little in the report as possible.

The line of defence should always be selected on the ground and not on the map.

The sketch is of great assistance in organizing the defence, but it does not take the place of an inspection of the ground. It is impossible for any sketch, for instance, to show how much crops and trees obscure the view.

10*b*. RECONNAISSANCE OF A POSITION OCCUPIED BY THE ENEMY.

In this the most that can usually be done is to make a rough sketch,—as an example study the reconnaissance of the ground across the Tugela, made with plane-table and range-finder, pp. 77, 78—mark on it what entrenchments, guns, camps, can be detected, and add in the form of a note any information which may elucidate the sketch, such as the nature of the ground,the enemy's flanks rest on, the existence of advanced posts, the condition of the ground as affecting the movement of all arms, the approaches leading to the position.

The following points should be ascertained as far as possible :—

    i. The extent of the position.
    ii. The weak parts of the position.
    iii. Any portions of the position which can be said to constitute "keys."
    iv. The best line of attack, and the physical features of which the possession will favour the development of an effective fire against the weak parts of the position.

In all reconnaissances of a position and of the ground over which troops are ordered to attack, every locality from which a covering fire can be maintained should be carefully noted.

11. RECONNAISSANCE SKETCH OF AN AREA,
otherwise a topographical sketch of an area with notes.

It will have been noticed by any officer who has read through this chapter, that much of the information required on any reconnaissance can be obtained from a good topographical map. For instance, on the 1-inch Ordnance Survey maps of the United Kingdom can be found—

    The heights and slopes of hills.
    The classes of roads.
    Railways, whether single or double; stations.

The position of smithies, post and telegraph offices.
The shape, position, and nature of woods.
Landmarks, such as churches and windmills.
The size of villages and towns, the positions of isolated houses and farms.
The existence of defiles and passes.
Rivers, streams, and watercourses; bridges; locks and weirs fords.
The nature of the coast line.

Again, for use in identifying one s position or finding one's way a topographical map is very superior to any linear sketch.

It follows from this that whenever good topographical maps are available, they should be freely used, and tracings or enlargements taken from them for all sorts of reconnaissances; for instance 2 inch enlargements from the 1-inch Ordnance map are very useful for showing the disposition of outposts or the organization of a defensive position.

It also follows that whenever a force is camped for any length of time in an unmapped country, the officer commanding the force should cause topographical area sketches of the neighbouring country to be made. This was done to a certain extent in China in 1901, where officers commanding posts were ordered to have sketches on the 1-inch scale made of the country within a radius of 10 miles of their posts. In the south African War of 1899-1902, in addition to the work of the Survey Sections, a great many sketches were produced by various officers, and sent for compilation to Pretoria, especially during the latter stages of the war. Nearly half a million sheets of compilations were issued to the force during the war.

In the Burmese wars of 1885-89 (excluding the valuable work of the survey of India) much compilation was carried out in Mandalay from material furnished by officers in posts and on column; but a good deal of this was poor stuff, as the plane-table was not sufficiently used. (In South Africa some officers had improvised plane-tables constructed.)

It is to be impressed on officers that the most generally useful sketch is the sketch of an area, which should approximate as far as time allows to the standard of a topographical map. Even a road sketch can be made far more intelligible by showing as much as possible of the surrounding country. Of course this principle must give way to the necessities of individual cases ; but a linear sketch (a road, route, river, or railway sketch), is, as a rule, useful only as a diagram to illustrate a report, and it is clear that the sketch of a defensive position, or of an outpost position, or of woods, villages, and farms, and of defiles and passes, must be of the nature of a rough topographical map. The various examples of topographical sketches given in this book should therefore be carefully studied, and the methods of constructing them should be mastered.

APPENDIX.

APPROXIMATE METHODS OF FINDING THE TRUE NORTH.

i. With a compass, knowing the magnetic variation, which can be obtained approximately from Plate XIII.
ii. In the northern hemisphere, in ordinary latitudes, the true bearing of the Pole Star is always within 2° of north.
iii. In the southern hemisphere the Southern Cross is approximately south when the long limb of the cross is vertical. An idea of the true south at other times may be obtained by remembering that this constellation is about 30° distant from the southern celestial pole.
iv. At the equinoxes (end of March and end of September) the sun rises due east and sets due west.
v. At apparent noon the sun is on the meridian. Apparent noon may, however, differ from mean noon (12 o'clock by a watch) by as much as 16 minutes. At noon by the watch the sun may be many degrees off the meridian.
vi. Outside the tropics a very rough approximation to the meridian may be found as follows: hold a watch face upwards and (in the northern hemisphere) point the hour hand at the sun, the line bisecting the angle at the centre of the watch between the hour hand and XII, is approximately in the meridian. In the southern hemisphere point XII at the sun. This rule is very rough.
vii. Hang a plumb bob from the top of an inclined stick which is stuck into the ground. Draw circles on the ground with the point where the plumb bob touches the ground as centre. Mark where the shadow of the top of the stick, cast by the sun, touches these circles. The middle point on any circle between the morning and afternoon shadow points is on the meridian of the plumb-bob.

NOTE.—None of these methods are to be used for determining the magnetic declination.

For accurate methods of finding the meridian, see Text Book of Topography.

Plate XX.   *To face p.* 116.

PLAN OF THE BRITISH FORT AT KOSHEH FOUND IN THE CAMP OF THE ENEMY AFTER THE BATTLE OF GINISS.

(Translation.)

A This is the fort of the infidels, the enemies of God, the liars, God curse them.
B This is the zeriba of wire, and after it the zeriba of thorns.
C The long gun.
D The mountain gun.
E The big gun.
F This is the place where they go down from the fort to drink water.
G The small steamer.
H The large steamer.
J The mountain gun.
K The rampart.
L The outside fort.
M The inside fort.

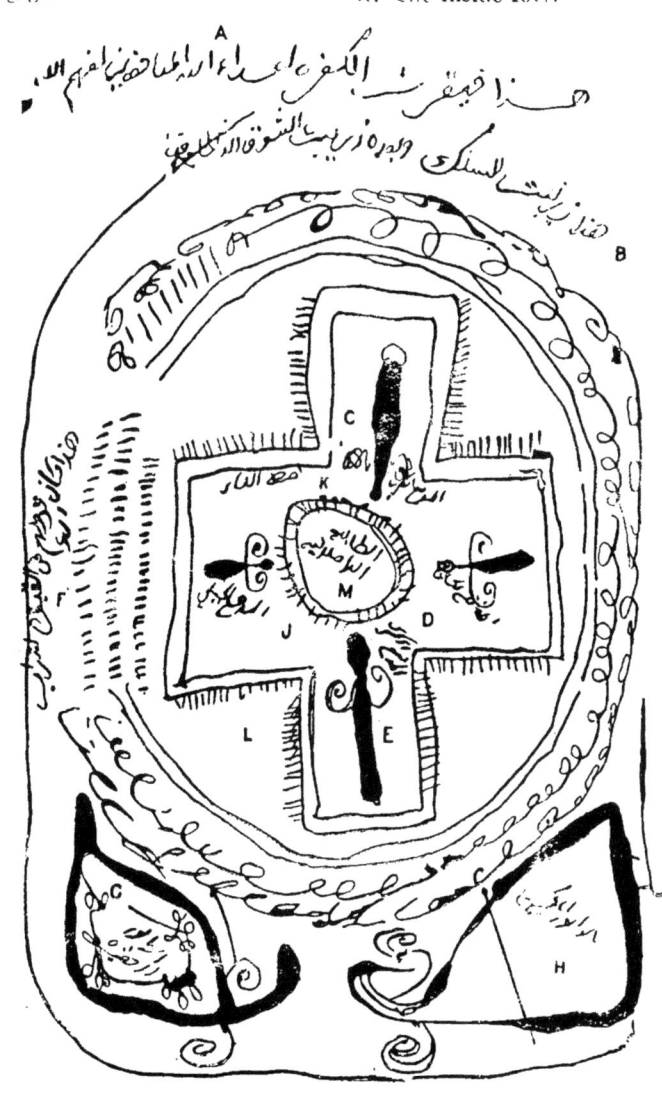

INDEX.

	PAGE
Abattis, conventional sign for	25
Abney's level	36
Accommodation for troops	105
,, reconnaissance for	105
Aneroid barometer	37, 43
Angle	9
Barometer, aneroid	37, 43
Base, or base line	9, 41, 59
Basin, definition of	8
Batteries, conventional sign for	25
Bearing, true and magnetic	9
,, forward	66
Bicycle, for traversing	69
Billeting	105
Bridges, conventional sign for	24
Burma, 1885 campaign	114
Campaigns quoted, *see* Wars.	
Camps, size of	99
Cavalry sketching board	62, 70
,, ,, scales for	73
Chain column	64
China campaign, 1901	13, 46, 114
Chitral ,, 1895	90
Coast reconnaissance	105
Col, definition of	8
Colours, conventional	24
Compass, prismatic	50
,, trough	39
,, sketching with	55
,, variation of	52
,, bearing, night march	84
Contours, definition of	9
,, advantages of	28
,, normal system of	32
Contouring on a plane-table	47
Conventional signs	23
Copying maps	19
Crest, definition of	8

	PAGE
Datum, definition of	9
Declination, magnetic	52
Defile, reconnaissance of	109
Demolitions, conventional sign for	25
Detail to be shown on 2-inch scale	61
Dip of needle	54
Direction lines	65
Dune, definition of	8
Enlarging maps	19
Entanglements, conventional sign	25
Escarpment, definition of	8
Eye-sketching	80
Fall, definition of	10
Field-book	10, 64
Fords, notes on	102
Forest, sketching in	65
Form-lines	10, 28, 30
Forward bearing	66
Frozen rivers	103
Gorge, definition of	8
Gradient ,,	10
,, road	99
Hachures	10, 27, 33
Hay, weight of	107
Heights, when plane-tabling	45
Hill features, representation of	27
,, shading	27
Horizontal equivalent	10, 30
,, hachures	27
Horseback, sketching on	70
Houses, conventional sign for	61
Instruments used in field sketching	35
Intersection on a plane-table	41

	PAGE		PAGE
Intrenchments, conventional sign	25	Plane-table resection on	43
		,, traversing with	46
Kilometre	95	Plateau, definition of	8
		Plotting, ,,	10
Landscape sketches	79	,, a traverse	63, 68
Layer system	28	Plumb-bob, for finding meridian	116
Lettering	25	Points of the compass	52
Luminous compass	50, 84	Pole Star	116
,, maps	84	Position, reconnaissance of	111
		Prismatic compass	50
Magnetic attraction, local	10, 54	,, ,, sketching	55
,, bearing	9	Protractor, use of	18, 55
,, declination	10, 52	Public house, sign for	24
,, meridian	10		
,, north point	54	Railway, conventional sign for	23
,, variation	10, 52	,, reconnaissance	103
Map reading	87	Range-finders	73
Map, definition	6	,, sketching with	77
,, luminous	84	Reconnaissance	90
,, ordnance 23, 27, 28, 32, 95, 113		,, for accommoda-	
,, their uses	5	tion	105
Marsh, conventional sign for	24	Reconnaissance of an area	113
Mekometer	73	,, coast	105
Meridian, definition of	10	,, of defiles and	
,, determination of	116	passes	109
Mounting a plane-table	39	Reconnaissance of outpost posi-	
		tions	110
Night marches	83	Reconnaissance of a position	111
North, methods of finding	116	,, of a railway	103
		,, reports	92
Offset, definition of	10, 63	,, of a river	100
,, column	65	,, road	96
Ordnance Maps 23, 27, 28, 32, 95, 113		,, route	100
Orienting, definition of	10	,, scales for	94
Outline	7	,, sketches	93
Outpost position, reconnaissance		,, for supplies	105
of	110	,, training in	94
Outpost duty, sketch by officer on	111	,, of villages and	
		farms	108
Pass, definition of	8	Reconnaissance of woods	107
,, reconnaissance of	109	Reducing maps	19
Pins, not to be used	40	Representative fraction	14
Plane-table	38	Resection	10, 43, 58
,, compared with com-		River, conventional sign for	24
pass	35, 54	,, fords	102
Plane-table contouring on	47	,, frozen	103
,, mounting a	39	,, reconnaissance 68, 100, 103	
,, surveying with	40	,, velocity of	102

	PAGE
Roads, classes of..	23
Saddle, definition of	9
Scales, definition of	12
,, comparative	16
,, diagonal..	17
,, plain	14
,, vernier ..	37
,, used in, South Africa, India, &c.	13
,, necessity of small scales	12
Sections ..	10
Setting a map	10, 88
,, cavalry sketching board	70
Sight-rule	39
Sketch, military, definition of ..	6
,, landscape	79
,, the uses of a	5
Sketching case ..	55
,, on horseback..	70
,, without instruments	80
Slope, degree of ..	30
South African War 6, 13, 33, 38, 68, 78, 114	
Southern cross, for finding meridian	116
Spur, definition of	9
Squares for enlarging, &c., maps	21
Stack, contents of	107
Stars, for marching by at night	86
Stations of a traverse ..	64
Straw, weight of..	107
Supplies, reconnaissance of	103
Telegraph office, conventional sign for	24
Topographical maps, definition	6

	PAGE
Topographical maps, contours on	28
,, ,, uses of	5, 113
Traverse, definition	10, 63
,, with plane-table	47
,, with field-book	63
,, when to use ..	63
Triangle of error	10, 43
Triangulation	11
Troops, conventional sign for ..	25
Trough-compass..	39
Variation, magnetic annual	52
,, ,, positional ..	52
,, ,, diurnal	52
,, of barometer..	38
Velocity of rivers	102
Vertical hachures	27, 33
,, interval, definition of	11, 30
Villages, reconnaissance of	107
Wars quoted—	
Burma, 1885 ..	114
China, 1901 ..	13, 114
Chitral, 1895 ..	90
Indian ..	13
Nile, 1885	6, 100
Sikkim, 1889 ..	6
South Africa, 1899–1902 6, 13, 33, 38, 68, 78, 100, 114	
Watch, use of, to find the meridian	116
Watercourse, definition of	9
Watershed	9
Water supply ..	107
Width of river, to find ..	103
Woods, reconnaissance of	107

LONDON:
PRINTED FOR HIS MAJESTY'S STATIONERY OFFICE,
BY HARRISON AND SONS, ST. MARTIN'S LANE,
PRINTERS IN ORDINARY TO HIS MAJESTY.

(Wt. 23446 3000 1 | 04—H & S 3845) $\frac{P\ 02}{847}$

PLEASE DO NOT REMOVE
CARDS OR SLIPS FROM THIS POCKET

UNIVERSITY OF TORONTO LIBRARY